Account Of A Tour In Normandy

Vol. II

By

Dawson Turner

Double9 BOOKS

Account Of A Tour In Normandy
Vol. II
by Dawson Turner

ISBN: 978-93-59957-16-6
Published by

DOUBLE 9 BOOKS
2/13-B, Ansari Road
Daryaganj, New Delhi – 110002
info@double9books.com
www.double9books.com
Tel. 011-40042856

ABOUT THE AUTHOR

Dawson Turner FRS FSA was an English banker, botanist, and antiquary who lived from 18 October 1775 to 21 June 1858. He specialized in cryptogam botany and was the father-in-law of botanist William Jackson Hooker and historian Francis Palgrave. Turner was the son of James Turner, the president of the Gurney and Turner's Yarmouth Bank, and Elizabeth Cotman, the only daughter of Yarmouth Mayor John Cotman. He attended North Walsham Grammar School (now Paston College) in Norfolk and Barton Bendish as a botanist's student. He then went to Pembroke College, Cambridge, where his uncle, Rev. Joseph Turner, was the Master. He did not graduate, however, due to his father's grave sickness. He joined his father's bank in 1796. In his spare time after becoming a banker, he became more interested in botany, collecting specimens in the field. Turner promised to assist James Sowerby with specimens in 1794. Turner wrote several books and collaborated with other botanists. He was elected a Fellow of the Royal Society in December 1802. He was named a foreign member of the Royal Swedish Academy of Sciences in 1816. He met Captain George Manby, an amateur artist, inventor, and Yarmouth barrack-master, through his first wife Mary.

CONTENTS

LETTER XIV
DUCLER—ST. GEORGES DE BOCHERVILLE—M. LANGLOIS

(*Ducler, July,* 1818.)

You will look in vain for Ducler in the *livre des postes*; yet this little town, which is out of the common road of the traveller, becomes an interesting station to the antiquary, it being situated nearly mid-way between two of the most important remains of ancient ecclesiastical architecture in Normandy—the abbeys of St. Georges de Bocherville and of Jumieges.—The accommodation afforded by the inns at Bocherville and Jumieges, is but a poor substitute for the hospitality of the suppressed abbeys; and, as even the antiquary must eat and perhaps sleep, he who visits either St. George or the holy Virgin, will do well to take his *fricandeau* and his bed, at the place whence I am writing.

At a period when the right bank of the Seine from Harfleur to Rouen displayed an almost uninterrupted line or monastic buildings, Ducler also boasted of a convent[1], which must have been of some importance, as early as the middle of the seventh century.—King Childeric IInd, granted the forest of Jumieges to the convent of the same name and that of St. Vandrille; and St. Ouen was directed by the monarch to divide the endowment between the two foundations. His award did not give satisfaction to St. Philibert, the abbot of Jumieges, who maintained that his house had not received a fair allotment. The proposition was stoutly resisted by St. Lambert, abbot of St. Vandrille; and the dispute was at length settled by the saints withdrawing their claims, and ceding the surplus land to the abbey of Ducler. St. Denys was the patron of this abbey; and to him also the present parochial church is dedicated: it is of Norman architecture; the tower is surrounded by a row of fantastic corbels; and a considerable quantity of painted glass yet remains in the windows. The village itself (for it is nothing more than a village, though honored by French geographers with the name of a bourg), consists of a single row of houses, placed immediately under the steep chalk cliff which borders the Seine. The face of the cliff is also indented by excavations, in which the poorer inhabitants dwell, almost like the Troglodytes of old. The situation of Ducler, and that of the two neighboring abbeys, is delightful in summer and in fine weather. In winter it must be cold and cheerless; for, besides being close to a river of so great breadth, it looks upon a flat marshy shore, whence exhalations copiously arise. The view from our chamber window this morning presented volumes of mist rolling on with the stream. The tide was setting in fast downwards; and the water glided along in silent rapidity, involved in clouds.

The village of Bocherville, or, as it is more commonly called, of St. Georges, the place borrowing its name from the patron saint of the abbey, lies, at the distance of about two leagues from Rouen. The road is exceedingly pleasing. Every turning presents a fresh view of the river; while, on looking back, the city itself is added to the landscape; and, as we approach, the abbey-church is seen towering upon the eminence which it commands.

The church of St. Georges de Bocherville, called in old charters *de Baucherville,* and in Latin *de Balcheri* or *Baucheri villa,* was built by Ralph de Tancarville, the preceptor of the Conqueror in his youth, and his chamberlain in his maturer age. The descendants of the founder were long the patrons and advocates of the monastery. The Tancarvilles, names illustrious in Norman, no less than in English, story, continued during many centuries to regard it as under their particular protection: they enriched it with their donations whilst alive, and they selected it as the spot to contain their remains when they should be no more.

The following portion of the charter, which puts us in possession of the indisputable æra of the erection of the church, is preserved by Mabillon[2]. It is the Conqueror who speaks.—"Radulfus, meus magister, aulæque et cameræ princeps, instinctu divino tactus, ecclesiam supradicti martyris Georgii, quæ erat parva, re-edificare a fundamentis inchoavit, et ex proprio in modum crucis consummavit."

The Monarch and his Queen condescended to gratify a faithful and favorite servant, by endowing his establishment. The corpse of the sovereign himself was also brought hither from St. Gervais, by the monks and clergy, in solemn procession, before it was carried to Caen[3] for interment.

Ralph de Tancarville, however, was not fortunate in the selection of the inmates whom he planted in his monastery. His son, in the reign of Henry Ist, dismissed the canons for whom it was first founded, and replaced them by a colony of monks from St. Evroul. Ordericus Vitalis, himself of the fraternity of St. Evroul, commemorates and of course praises the fact. Such changes are of frequent occurrence in ecclesiastical history; and the apprehension of being rejected from an opulent and well-endowed establishment, may occasionally have contributed, by the warning example, to correct the irregularities of other communities. A century later, the abbot of St. Georges was compelled to appeal to the pope, in consequence of an attempt on the part of his brethren at St. Evroul, to degrade his convent into a mere cell, dependent upon theirs.—The chronicle of the abbey is barren of events of general interest; nor do its thirty-one abbots appear to have been men of whom there was much more to be said, than that they arrived at their dignity on such a year, and quitted it on such another. Of the monks, we are told that, in the fifteenth century, though their number was only eight, the dignitaries included, the daily task allotted them was greater than would in any of the most rigid establishments, in latter days, have been imposed upon forty brethren in a week!

Inconsiderable as is the abbey, in an historical point of view, the church of St. Georges de Bocherville is of singular importance, inasmuch as it is one of the land-marks of Norman architecture. William, in his charter, simply styles himself Dux Normannorum; it therefore was granted a few years before the conquest. The building has suffered little, either from the hands of the destroyers, or of those who do still more mischief, the repairers; and it is certainly at once the most genuine and the most magnificent specimen of the circular style, now existing in Upper Normandy.—The west front is wholly of the time of the founder, with the exception of the upper portion of the towers that flank it on either side. In these are windows of nearly the earliest pointed style; and they are probably of the same date as the chapter-house, which was built in the latter part of the twelfth century. The effect of the front is imposing: its general simplicity contrasts well with the rich ornaments of the arched door-way, which is divided into five systems of mouldings, all highly wrought, and presenting almost every pattern commonly found in Norman buildings. A label encircles the whole, the inner edge of which is indented into obtuse pyramids, erroneously called lozenges. The capitals of the columns supporting the arch are curiously sculptured: upon the second to the left, on entering, are Adam and Eve, in the act of eating the forbidden fruit; upon the opposite one, is represented the Flight into Egypt. Normandy does not contain, I believe, a richer arch; but very many indeed are to be seen in England, even in our village churches, superior in decoration, though not, perhaps, in size; for this at St. Georges is on a very large scale: on each side of it is a smaller blank arch, with a single moulding and a single pillar. Two tiers of circular-headed windows of equal size fill up the front.—The rest of the exterior may be said to be precisely as it was left by the original builders, excepting only the insertion of a pointed window near the central tower.

The inside is at least equally free from modern alterations or improvements. No other change whatever is to be traced in it than such as were required to repair the injuries done it during the religious wars; and these were wholly confined to a portion of the roof, and of the upper part of the wall on the south side of the nave. The groined roof, though posterior to the original date of the building, is perhaps of the thirteenth century. The nave itself terminates towards the east in a semi-circular apsis, according to the custom of the times; and there, as well as at the opposite extremity of the building, it has a double tier of windows, and has columns more massy than those in the body of the church. The aisles end in straight lines; but, within, a recess is made in the thickness of the wall, for the purpose of admitting an altar. Both the transepts are divided within the church, at a short distance from their extremities, into two stories, by a vaulted roof of the same height as the triforium.—M. Le Prevost, who has very kindly communicated to me the principal part of these details, has observed the same to be the case in some other contemporary buildings in Normandy. On the eastern side of each transept is a small chapel, ending, like the choir, in a semi-circular apsis, which rises no higher than the top of the basement story. A cable moulding runs round the walls of the whole church within.—You and I, in our own country, have often joined in admiring the massy grandeur of Norman architecture, exemplified in the nave of Norwich cathedral: at St. Georges I was still more impressed by the noble effect of semi-circular arcades, seen as they are here on a still larger scale, and in their primitive state, uninterrupted and undebased by subsequent additions.

On closer examination, the barbarous style of the sculpture forces itself upon the eye. Towards the western end of the building the capitals are comparatively plain: they become more elaborate on approaching the choir. Some of them are imitations or modifications (and it may even be said beautiful ones) of the Grecian model; but in general they are strangely grotesque. Many represent quadrupeds, or dragons, or birds, and commonly with two bodies, and a single head attached to any part rather than the neck. On others is seen "the human form divine," here praying, there fighting; here devouring, there in the act of being devoured; not uncommonly too the men, if men they must be called, are disfigured by enormous heads with great flapping ears, or loll out an endless length of tongue.—One is almost led to conceive that Schedel, the compiler of the *Nuremberg Chronicle*, had a set of Norman capitals before his eyes, when he published his inimitable series of monsters. His "homines cynocephali," and others with "aures tam magnas ut totum corpus contegant," and those again whose under lips serve them as coverlids, may all find their prototypes, or nearly so, in the carvings of St. Georges.

The most curious sculptures, however, in the church, are two square bas-reliefs, opposite to one another, upon the spandrils of the arches, in the walls that divide the extremities of the transepts into different stories[4].

They are cut out of the solid stone, in the same manner as the subjects on the block of a wood-engraving: one of these tablets represents a prelate holding a crosier in his left hand, while the two fore-fingers of the right are elevated in the act of giving the blessing; the other contains two knights on horseback, jousting at a tournament. They are armed with lance and buckler, and each of them has his head covered with a pointed helmet, which terminates below in a nasal, like the figures upon the Bayeux tapestry.—This coincidence is interesting, as deciding a point of some moment towards establishing the antiquity of that celebrated relic, by setting it beyond a doubt that such helmets were used anterior to the conquest; for it is certain that these basso-relievos are coeval with the building which contains them.

This church affords admirable subjects for the pencil. It should be drawn in every part: all is entire; all original; the corbel-stones that support the cornice on the exterior are perfect, as well along the choir and nave, as upon the square central steeple: each of the sides of this latter is ornamented with a double tier of circular arches. The buttresses to the church are, like those of the chapel of St. Julien, shallow and unbroken; and they are ranged, as there, between the windows. At the east end alone they take the shape of small semi-cylindrical columns of disproportionate length.

Capital, in the Chapter House at St. Georges.

The monastic buildings, which were probably erected about the year 1700, now serve as a manufactory. Between them and the church is situated the chapter-house, which was built towards the end of the twelfth century, at a period when the pointed architecture had already begun to take place of the circular style. Its date is supplied in the *Gallia Christiana*, where we read, that Victor, the second abbot, "obiit longævus dierum, idibus Martii, seu XVIII calendas Aprilis, ante annum 1211; sepultusque est sub tabulâ marmoreâ in capitulo quod erexerat."

We found it in a most ruinous and dilapidated state, yet extremely curious; indeed not less so than the church. Its front to the west exhibits a row of three semi-circular arches, with an ornament on the archivolt altogether different from what I recollect to have seen elsewhere[5]. The inside corresponds in profuse decoration with this entrance; but the arches in it are all pointed. An entablature of beautiful workmanship is carried round the whole building, which is now used as a mill: it was crowded with dirty children belonging to the manufactory; and the confusion which prevailed, was far from being favorable to the quiet lucubrations of an antiquary. In no part of the church is the sculpture equally curious; and it is very interesting to observe the progress which this branch of the art had made in so short a time. Two or three of the capitals to the arches in front, seem to include one continued action, taken apparently from the history of Joshua. Another capital, of which I send you a sketch from the pencil of M. Le Prevost, is a great curiosity. The group which it contains, is nearly a duplicate of the supposed statue of William the Conqueror at Caen. In all probability it represents some legendary story, though the subject is not satisfactorily ascertained. Against the pillars that support these arches, were affixed whole-length figures, or cariatides, in alto-relievo. Three of them still remain, though much mutilated; two women and a man. They hold in their hands labels, with inscriptions that fall down to their feet in front. One of the females has her hair disposed in long braided tresses, which reach on either side to her girdle. In this respect, as well as in the style of the sculpture and costume, there is a resemblance between these statues and those on the portals at St. Denys and at Chartres, as well as those formerly on that of St. Germain des Prés, at Paris, all which are figured by Montfaucon in his Monumens de la Monarchie Française, and are supposed by him to be of the times of the Merovingian or Carlovingian dynasty; but subsequent writers have referred them to the eleventh or twelfth century.

It was in this chapter-house that M. Langlois[6] found, among a heap of stones, a most interesting capital, that had formerly been attached to a double column. By his kindness, I inclose you two drawings of it. One of them shews it in its entire form as a capital; the other exhibits the bas-relief carved upon it[7].

The various injuries sustained by the building, render it impossible to ascertain the spot which this capital originally occupied; but M. Le Prevost supposes that it belonged to some gate of the cloister, which is now destroyed. A more curious series of musical instruments is, perhaps, no where to be found; and it is a subject upon which authors in general are peculiarly unsatisfactory. I am told that, in an old French romance, the names of upwards of twenty are enumerated, whose forms and nature are quite unknown at the present day; while, on the other hand, we are all of us aware that painting and sculpture supply figures of many, for which it would be extremely difficult or impossible to find names[8].

The chapter-house, previously to the revolution, contained a tomb-stone[9], uninscribed and exhibiting only a sculptured sword, under which it was supposed that either Ralph de Tancarville himself, the founder of the abbey, or his grandson, William, lay interred. It is of the latter that the records of the monastery tell, how, on the fifth day after he girded himself with the military belt, he came to the church, and deposited his sword upon the altar, and subsequently redeemed it by various donations, and by confirming to the monks their right to the several benefices in his domain, which had been ceded to them by his grandfather.—Here then, I quit you: in a few days I shall have paid my devotions at the shrine of Jumieges:—meanwhile, in the language of the writers of the elder day, I close this sheet with.

EXPLICIT FELICITER Stus. GEORGIUS DE BOCHERVILLA;

DEO GRATIAS.

Footnotes:

[1] Histoire de la Haute Normandie, II. p. 266. VOL. II.

[2] Ann. Benedict. III. p. 674, 675.—This charter was not among the archives of the monastery; but I am informed by M. Le Prevost, that several are still in existence, most of them granted by the family of the founder, but some by Kings of England. One of the latter is by Richard Coeur de Lion, and his seal of red wax still remains appended to it, in fine preservation. The seal, on one side, represents the king seated upon his throne, with a pointed beard, having his crown on his head, and a sword in one hand, and sceptre in the other: on the other side, he is on horseback, with his head covered with a cylindrical helmet, surmounted with a very remarkable crest, in the form of a fan: on his shield are plainly distinguishable the three lions of England.—From among the charters granted by the Tancarville family, M. Le Prevost has sent me copies of two which have never yet been printed; but which appear to deserve insertion here. One is from Lucy, daughter of William de Tancarville, and grand-daughter of Ralph, the chamberlain.—"Notum sit Ricardo de Vernon and Willelmo Camerario de Tancarvilla, et veteribus et juvenibus, quòd Lucia, filia Willelmi, Camerarii de Tancarvilla, pro animâ suâ et pro animabus antecessorum suorum, ad ecclesiam Sti. Georgii de Bauchervilla dedit molendinum de Waldinivilla, quod est subter aliud molendinum et molendinum de Waldinval, liberè et quietè, et insupèr ecclesiam de Seonvilla, salvâ elemosinâ

Roberti sacerdotis in vitâ suâ, si dignus est habendi eam. Et post mortem Willelmi capellani sui de Sancto Flocello, ad ecclesiam suprà dictam dedit decimam de vavassoribus de Seolvilla, quam dedit in elemosinâ habendam Willelmo capellano totâ vitâ bene et in pace et securè, et decimas de custodiis totius terre sue que est in Constantino.—Ego Lucia do hanc elemosinam pro animâ meâ et pro antecessoribus ad ecclesiam Sanctii Georgii; et qui auferet ab eâ et auferetur ab eo regnum Dei. Amen.—Testibus, Ricardo de Haia et Matille uxore suâ et Nigello de Chetilivilla et hominibus de Sancto Flocello."—To this is added, in a smaller hand-writing, probably the lady's own autograph, the following sentence:—"Et precor vos quòd ecclesia Sancti Georgii non decrescatur in tempore vestro pro Dei amore et meo de elemosinis patris mei neque de meis."—There is still farther subjoined, in a different hand-writing, and in a much paler ink:—"Hæc omnia Ricardus de Vernon libenter concessit."—The other charter was granted by William the Younger, and details a curious custom occasionally observed in the middle ages, in making donations:—

"Universis sancte ecclesie fidelibus. Willelmus junior camerarius in domino salutem. Notum sit presentibus et futuris, quod ego Willelmus junior camerarius quinto die post susceptum militie cingulum veni apud Sanctum Georgium, ibique cum honorificâ processione suscepérunt me Abbas Ludovicus et monachi cum magno gaudio letantes; et ibi obtuli gladium meum super altare Sti. Georgii, et tunc consilio et admonitione sociorum meorum nobilium virorum qui mecum venerant, scilicet Roberti des Is, dapiferi mei, et Rogerii de Calli, et Johannis de Lunda, et aliorum plurium, redemi gladium meum per dona et confirmationem plurium ecclesiarum, quas ipso die concessi eisdem meo dono, et, sicut avus meus, fundator illius monasterii dederat, confirmavi; scilicet ecclesiam de Abetot et ecclesiam de Espretot cum decimâ, et ecclesiam Sancti Romani cum duabus partibus decime, et similitèr ecclesiam de Tibermaisnil: confirmavi etiam dona militum meorum et amicorum quæ dederunt ipso die abbatie in perpetuam elemosynam, Rogerius de Calli dedit XX Sot. annuatìm; Robertus de Mortùomari X Sot.; Robertus des Is X solidos; Johannes de Lunda, cognatus meus X Sot.; Andreas de Bosemuneel X solidos, vel decimam de una carrucatura terre ... Humfridus de Willerio X solid.; Willelmus de Bodevilla X acras terre; Garinus de Mois V solid.; Adam de Mirevilla X solid.; Robert. de Fuschennis X solid.; Lesra de Drumara I acram terre."

[3] The following are the words of Ordericus Vitalis, upon the subject:

"Religiosi tandem viri, Clerici et Monachi, collectis viribus et intimis sensibus, processionem ordinaverunt: honestè induti, crucibus et thuribus, ad Sanctum Georgium processerunt, et animam Regis, secundum morem sanctæ Christianitatis Deo commendaverunt."—*Duchesne, Scriptores Normanni,* p. 661.

[4] See Cotman's Architectural Antiquities of Normandy, t. 10. f. A. and B.

[5] See Cotman's Architectural Antiquities of Normandy, t. 11. last figure.

[6] My readers will join with me, I trust, in thanks to M. Langlois, for his drawings; and will not be sorry to see, accompanying his sketch of the bas-relief, a spirited one of himself. Normandy does not contain a more ardent admirer of her antiquities, or one to whom she is more indebted for investigating, drawing, and publishing them. But, to the disgrace of Rouen, his labors are not rewarded. All the obstacles, however opposed by the "durum, pauperies, opprobium," have not been able to check his independent mind: he holds on his course in the illustration of the true Norman remains; and to any antiquary who visits this country, I can promise a great pleasure in the examination of his port-folio.

[7] Its size at top is fourteen inches and a half, by six inches and two-thirds.

[8] This difficulty, in the present instance, has yielded to the extensive researches of Mr. Douce, who has afforded

assistance to me, which, perhaps, no other antiquary could have bestowed. He has unravelled all the mysteries of minstrelsy with his usual ability; and I give the information in his own words, only observing that the numbers begin from the left.—"No. 1 was called the violl, corresponding with our Viol de Gamba. As this was a larger violin, though the sculptor has not duly expressed its comparative bulk, I conceive it was either used as a tenor or base, being perfectly satisfied, in spite of certain doubts on the subject, that counterpoint was known in the middle ages.—No. 2 is the largest instrument of the kind that I have ever seen, and it seems correctly given, from one part of it resting on the figure, No. 3, to support it. Twiss mentions one that he saw sculptured on the cathedral, at Toro, five feet long. The proper name of it is the rote, so called from the internal wheel or cylinder, turned by a winch, which caused the bourdon, whilst the performer stopped the notes on the strings with his fingers. This instrument has been very ignorantly termed a vielle, and yet continues to be so called in France. It is the modern Savoyard hurdy-gurdy, as we still more improperly term it; for the hurdy-gurdy is quite a different instrument. In later times, the rote appears to have lost its rank in concert, and was called the beggar's lyre.—No. 4 is evidently the syrinx, or Pan's pipe, which has been revived with so much success in the streets of London.—Twiss shewed me one forty years ago, that he got in the south of France, where they were then very common.—No. 5 is an instrument for which I can find no name, nor can I immediately call to memory any other representation of it. It has some resemblance to the old Welsh fiddle or crowth; but, as a bow is wanting, it must have been played with the fingers; and I think the performer's left hand in the sculpture does seem to be stopping the strings on the upper part, or neck, a portion of which has been probably broken off.—I suspect it to be the old mandore, whence the more modern mandolin. The rotundity of the sounding-board may warrant this conjecture.—No. 6 was called the psalterion, and is of very great antiquity, (I mean as to the middle ages).—Its form was very diversified, and frequently triangular. It was played with a plectrum, which the performer holds in his right hand.—No. 7 is the dulcimer, which is very common in sculpture. This instrument appears, as in the present case, to have been sometimes played with the fingers only, and sometimes with a plectrum.—No. 8 is the real vielle, or violin, of very common occurrence, and very ancient.—No. 9 is a female tumbler, or tomllesterre, as Chaucer calls them. This profession, so far as we can depend on ancient representation, appears to have exclusively belonged to women.—No. 10. A harp played with a plectrum, and, perhaps, also with the left hand occasionally.—No. 11. The figure before the suspended bells has had a hammer in each hand with which to strike them, and the opposite, and last, person, who plays in concert with him, has probably had a harp, as is the case in an ancient manuscript psalter illumination that I have, prefixed to the psalm Exaltate Deo.—I have seen these bells suspended (in illumination to the above psalm) to a very elegant Gothic frame, ascending like the upper part of a modern harp."

[9] Gallia Christiana, XI. p. 270.

Monastery of Jumieges

LETTER XV
ABBEY OF JUMIEGES—ITS HISTORY—ARCHITECTURAL DETAILS—TOMBS OF AGNES SOREL AND OF THE ENERVEZ

(*Ducler, July,* 1818)

The country between Ducler and Jumieges is of much the same character with that through which we had already travelled from Rouen; the road sometimes coasting the Seine, and sometimes passing through a well-wooded country, pleasantly intermingled with corn-fields. In its general appearance, this district bears a near resemblance to an English landscape; more so, indeed, than in any other part of Normandy, where the features of the scenery are upon a larger scale.

The lofty towers of the abbey of Jumieges are conspicuous from afar: the stone of which they are built is peculiarly white; and at a distance scarcely any signs of decay or dilapidation are visible. On a nearer approach, however, the Vandalism of the modern French appears in full activity. For the pitiful value of the materials, this noble edifice is doomed to destruction. The arched roof is beaten in; and the choir is nearly levelled with the ground. Two cart-loads of wrought stones were carried away, while we were there; and the workmen were busily employed in its demolition. The greater part, too, of the mischief, appears recent: the fractures of the walls are fresh and sharp; and the fresco-paintings are unchanged.—Had the proud, abbatial structure but been allowed to have existed as the parochial church of the village, the edifice might have stood for ages; but the French are miserably deficient in proper feeling; and neither the historical recollections connected with Jumieges, nor its importance as a monument of architectural antiquity, could redeem it from their tasteless selfishness. In a few years, its very ruins will have perished; and not a wreck will remain of this ancient sanctuary of religion and of learning.

It was in the year 654 or 655, that St. Philibert, second abbot of Rebais, in the diocese of Meaux, founded this monastery. He selected the site upon which the present building stands, a delightful situation, in a peninsula on the right bank of the Seine. This peninsula, and the territory extending from Ducler to Caudebec, had been granted to him for this purpose by Clovis

IInd, or, more properly speaking, by Bathilda, his queen; for the whole administration of affairs was in reality under her guidance, though the reins of state were nominally held by her feeble husband. The territory[10] had previously borne the name of Jumieges, or, in Latin, Gemeticum, a term whose origin has puzzled etymologists. Those who hold it disgraceful to be ever at a loss on points of this nature, and who prefer displaying a learned to an unlearned ignorance, derive Gemeticum, either from gemitus, because, "pro suis offensis illìc gemunt, qui in flammis ultricibus non erunt gemituri;" or from gemma, conformably to the following distich,—

"Gemmeticum siquidem a gemmâ dixere priores;

Quòd reliquis gemmæ, præcelleret instar Eoæ."

The ground upon which the abbey was erected was previously occupied by an ancient encampment. The author of the Life of St. Philibert, who mentions this circumstance, has also preserved a description of the original church. These authentic accounts of edifices of remote date, which frequently occur in hagiology, are of great value in the history of the arts[11].—The bounty of the queen was well employed by the saint; and the cruciform church, with chapels, and altars, and shrines, and oratories, on either side, and with its high altar hallowed by relics, and decked out with gold and silver and precious stones, shews how faithfully the catholics, in their religious edifices of the present day, have adhered to the models of the early, if not the primitive, ages of the church.

Writers of the same period record two facts in relation to Jumieges, which are of some interest as points of natural history.—Vines were then commonly cultivated in this place and neighborhood;—and fishes of so great a size, that we cannot but suppose they must have been whales, frequently came up the Seine, and were caught under the walls of the monastery.—The growth of the vine is abundantly proved: it is not only related by various monkish historians, one of whom, an anonymous writer, quoted by Mabillon, in the Acta Sanctorum ordinis Sancti Benedicti, says, speaking of Jumieges, "hinc vinearum abundant botryones, qui in turgentibus gemmis lucentes rutilant in Falernis;" but even a charter of so late a date as the year 1472, expressly terms a large tract of land belonging to the convent, the vineyard[12].—The existence of the English monastic vineyards has been much controverted, but not conclusively. Whether these instances of the northern growth of the vine, as a wine-making plant, do or do not bear upon the question of the supposed refrigeration of our climate by the increase of the Polar ice, must be left to the determination of others.—The whale-fishery of Jumieges rests upon the single authority of the Gesta Sancti Philiberti: the author admits, indeed, that it is a strange thing, "et a sæculo inauditum;" but still he speaks of it as a fact that has fallen under his own knowledge, that the monks, by means of hooks, nets, and boats, catch sea-fish[13], fifty feet in length, which at once supply their table with food, and their lamps with oil.

The number of holy men who originally accompanied St. Philibert to his new abbey, was only seventy; but they increased with surprising rapidity; insomuch, that his successor, St. Aicadras, who received the pastoral staff, after a lapse of little more than thirty years from the foundation of Jumieges, found himself at the head of nine hundred monks, besides fifteen hundred attendants and dependants of various denominations.

During all these early ages, the monastery of Jumieges continued to be accounted one of the most celebrated religious houses in France. Its abbots are repeatedly mentioned in history, as enjoying the confidence of sovereigns, and as charged with important missions. In their number, was Hugh, grandson of Pépin le Bref, or, according to other writers, of Charlemagne. Here also, Tassilo, Duke of Bavaria, and his son, Theodo, were compelled to immure themselves, after the emperor had deposed them; whilst Anstruda, daughter of Tassilo, was doomed to share his imperial bed.

An æra of misfortune began with the arrival of the Normans. It was in May, in the year 841, that these dreadful invaders first penetrated as far as Rouen, marking their track by devastation. On their retreat, which almost immediately succeeded, they set fire to Jumieges, as well as to the capital. In their second invasion, under Ironside and Hastings, the "fury of the Normans" was poured out upon Neustria; and, during their inroad, they levelled Jumieges with the ground[14]. But the monks saved themselves: they dispersed: one fled as far as St. Gall; others found shelter in the royal abbey of St. Denis; the greater part re-assembled in a domain of their own, called Haspres, in Flanders, whither they carried with them the bodies of St. Aicadrus and St. Hugh: there too they resided till the conversion of their enemies to Christianity.

The victorious fleet of Rollo first sailed in triumph up the Seine, in the year 876. According to three monkish historians, Dudo of St. Quintin, William of Jumieges, and Matthew of Westminster, the chieftain venerated the sanctity of Jumieges, and deposited in the chapel of St. Vast, the corpse of the holy virgin, Hameltruda, whom he had brought from Britain. They also tell us that, on the sixth day after his baptism, he made a donation of some lands to this monastery.—The details, however, of the circumstances connected with the first, diminish its credibility; and Jumieges, then desolate, could scarcely contain a community capable of accepting the donation. But under the reign of the son and successor of Rollo, the abbey of Jumieges once more rose from its ashes. Baldwin and Gundwin, two of the monks who had fled to Haspres, returned to explore the ruins of the abbey: they determined to seclude themselves amidst its fire-scathed walls, and to devote their lives to piety and toil.—In pursuing the deer, the Duke chanced to wander to Jumieges, and he there beheld the monks employed in clearing the ground. He listened with patience to their narration; but when they invited him to partake of their humble fare, barley-bread and water, he turned from them with disdain. It chanced, however, that immediately afterwards, he encountered in the forest a boar of enormous size. The beast unhorsed him, and he was in danger of death. The peril he regarded as a judgment from heaven; and, as an expiation for his folly, he rebuilt the monastery. So thoroughly, however, had the Normans demonachised Neustria, that William Longa Spatha was compelled to people the abbey with a colony from Poitou; and thence came twelve monks, headed by Abbot Martin, whom the duke installed in his office in the year 930. William himself also desired to take refuge from the fatigues of government in the retirement of the monastery; and though dissuaded by Abbot Martin, who reminded him that Richard, his infant, son still needed his care, he did not renounce his intention:—but his life and his reign were soon ended by treachery.

This second æra of the prosperity of Jumieges was extremely short; for the prefect, whom Louis d'Outremer, King of France, placed in command at Rouen, when he seized upon the young Duke Richard, pulled down the walls of this and of all the other monasteries on the banks of the Seine, to assist towards the reparation and embellishment of the seat of his government. But from that time forward the tide of monastic affairs flowed in one even course of prosperity; though the present abbatial church was not begun till the time of Abbot Robert, the second of that name, who was elected in 1037. By him the first stone of the foundation was laid, three years after his advancement to the dignity; but he held his office only till 1043, when Edward the Confessor invited him to England, and immediately afterwards promoted him to the Bishopric of London.—Godfrey, his successor at Jumieges, was a man conversant with architecture, and earnest in the promotion of learning. In purchasing books and in causing them to be transcribed, he spared neither pains nor expence. The records of the monastery contain a curious precept, in which he directs that prayers should be offered up annually upon a certain day, "pro illis qui dederunt et fecerunt libros."—The inmates of Jumieges continued, however, to increase in number; and the revenues of the abbey would not have been adequate to defray the expences of the new building, had not Abbot Robert, who, in 1050, had been translated to the see of Canterbury, supplied the deficiency by his munificence, and, as long as he continued to be an English prelate, remitted the surplus of his revenues to the Norman abbey. He held his archiepiscopal dignity only one year, at the expiration of which he was banished from England: he then retired to Jumieges, where he died the following spring, and was buried in the choir of the church which he had begun to raise. At his death, the church had neither nave nor windows; and the whole edifice was not completed till November, in the year 1066. In the following July the dedication took place. Maurilius, Archbishop of Rouen, officiated, in great pomp, assisted by all the prelates of the duchy; and William, then just returned from the conquest of England, honored the ceremony with his presence.

I have dwelt upon the early history of this monastery, because Normandy scarcely furnishes another of greater interest. In the *Neustria Pia*, Jumieges fills nearly seventy closely-printed folio pages of that curious and entertaining, though credulous, work.—What remains to be told of its annals is little more than a series of dates touching the erection of different parts of the building: these, however, are worth preserving, so long as any portion of the noble church is permitted to have existence, and so long as drawings and engravings continue to perpetuate the remembrance of its details.

The choir and extremities of the transept, all of pointed architecture, are supposed to have been rebuilt in 1278.—The Lady-Chapel was an addition of the year 1326.—The abbey suffered materially during the wars between England and France, in the reigns of our Henry IVth and Henry Vth: its situation exposed it to be repeatedly pillaged by the contending parties; and, were it not that the massy Norman architecture sufficiently indicates the true date, and that we know our neighbors' habit of applying large words to small matters, we might even infer that it was then destroyed as effectually as it had been by Ironside: the expression, "lamentabilitèr desolata, diffracta et annihilata," could scarcely convey any meaning short of utter ruin, except to the ears of one who had been told that a religious edifice was actually abimé during the revolution, though he saw it at the same moment standing before him, and apparently uninjured.—The arched roof of the choir received a complete repair in 1535: that of the nave, which was also in a very bad state, underwent the same process in 1688; at the same time, the slender columns that support the cornice were replaced with new ones, and the symbols of the Evangelists were inserted in the upper part of the walls. These reparations are managed with a singular perception of propriety; and though the manner of the sculpture in the symbolic figures, is not that of a Gothic artist, yet they are most appropriate, and harmonize admirably with the building.

You must excuse me that, now I am upon this subject, I venture to "travel somewhat out of the record," for the sake of proposing to you a difficulty which has long puzzled me:—the connection which Catholic divines find between St. Luke's Bull and the word Zecharias;—for it appears, by the following distich from the Rhenish Testament, that some such cause leads them to regard this symbol as peculiarly appropriate to the third Evangelist:—

> "Effigies vituli, Luca, tibi convenit; extat
> Zacariæ in scriptis mentio prima tuis."—

An antiquary might be perplexed by these figures, the drawings whereof I now send you. He would find it impossible to suppose the exquisitely-sculptured images and the slender shafts with richly-wrought capitals, of the same date as the solid simple piers and arches all around; and yet the stone is so entirely the same, and the workmanship is so well united, that it would require an experienced eye to trace the junction. In the middle of the sixteenth century, the central tower was also found to need reparation;

and the church, upon this occasion, sustained a lasting injury, in the loss of its original spire, which was of lead, and of great height and beauty. It was taken down, under pretence of its insecurity; but in reality the monks only wished to get the metal. This happened in 1557, under Gabriel le Veneur, Bishop of Evreux, the then abbot. Five years afterwards the ravages of the Huguenots succeeded: the injury done to Jumieges by these sectaries, was estimated at eighty thousand francs; and the library and records of the convent perished in the devastation.

The western front of the church still remains almost perfect; and it is most singular. It consists, of three distinct parts; the central division being nearly of equal width to the other two conjointly, and projecting considerably beyond them. The character of the whole is simplicity: the circular door-way is comparatively small, and entirely without ornament, except a pillar on each side; the six circular-headed windows over the entrance, disposed in a double row, are equally plain. Immediately above the upper tier of windows, is a projecting chequered cornice; and, still higher, where the gable assumes a triangular form, are three lancet-shaped apertures, so extremely narrow, that they resemble the loop-holes of a dungeon rather than the windows of a church. In each of the lateral compartments was likewise originally a door-way, and above it a single window, all of the same Norman style, but all now blocked up. These compartments are surmounted with short towers, capped with conical spires. The towers appear from their style and masonry to be nearly coeval with the lower part of the building, though not altogether so: the southern is somewhat the most modern. They are, however, so entirely dissimilar in plan from the rest of the front, that we cannot readily admit that they are a portion of the original design. Nor are they even like to each other. Both of them are square at their bases, and preserve this form to a sufficient height to admit of two tiers of narrow windows, separated from each other by little more than a simple string-course. Above these windows both become octagon, and continue so to the top; but in a very different manner. The northern one has obtuse angles, imperfectly defined; the southern has four projecting buttresses and four windows, alternating with each other. The form of the windows and their arrangement, afford farther marks of distinction. The octagon part is in both turrets longer than the square, but, like it, divided into two stories.

The central tower of the church, which was large and square, is now reduced to a fragment: three of its sides are gone; the western remains sufficiently perfect to shew what the whole was when entire. It contained a double tier of arches, the lower consisting of two, which were large and simple, the upper of three, divided by central shafts and masonry, so that each formed a double window. All of them were circular-headed, but so far differed from the architecture of the nave, that they had side-pillars with capitals.

The church[15] was entered by a long narrow porch.—The nave is a fine specimen of Norman architecture, but is remarkable in that style for one striking peculiarity, that the eight wide circular arches on either side, which separate it from the aisles, are alternately supported by round pillars and square piers; the latter having semi-cylindrical columns applied to each of their sides. The capitals are ornamented with rude volutes. The arches in the triforium are of nearly the same width as those below, but considerably less in height. There is no archivolt or moulding or ornament. Above these there is only one row of windows, which, like all the rest, are semi-circular headed; but they have neither angular pillars, nor mouldings, nor mullions. These windows are rather narrow externally, but within the opening enlarges considerably. The windows in the upper and lower tiers stand singly: in the intermediate row they are disposed by threes, the central one separated from the other two by a single column.—The inside of the nave is striking from its simplicity: it is wholly of the eleventh century, except the reparations already mentioned, which were made in 1688.—The choir and Lady-Chapel are nearly demolished; and only some fragments of them are now standing: they were of pointed architecture, and posterior to the nave by at least two centuries.

A smaller church, dedicated to St. Peter, stood near the principal one, with which it was connected by means of a corridor of pointed arches. There are other instances of two churches being erected within the precincts of one abbey, as at Bury St. Edmund's. St. Peter's was a building at least of equal antiquity with the great church. But it had undergone such alterations in the year 1334, during the prelacy of the twenty-seventh abbot, William Gemblet, that little of the original structure remained. He demolished nearly the whole of the nave, for the sake of adding uniformity to the cloisters of the monastery.—M. Le Prevost, however, is of opinion, that the ruins of Jumieges contain nothing more interesting to an antiquary than the west end of the portion of building, which subsequently served as the nave. It is a mass of flint-work; and he considers it as having belonged to the church that existed before the incursion of the Normans.

The cloisters, which stood to the south-west of St. Peter's, are now almost wholly destroyed.—To the west of them is a large hall or gallery, known by the name of la Salle des Chevaliers. It is entered by two porches, one towards the north-west, the other towards the south-west[16], both full of architectural beauty and curiosity. I know of no authority for their date; but, from the great variety and richness of their ornaments, and the elegant taste displayed in the arrangement of these, I should suppose them to have been erected during the latter half of the twelfth century: one of the arches is unquestionably pointed, though the cusp of the arch is very obtuse. The slight sketch which accompanies this letter, represents a fragment of the inner door-way of the south-west porch, and may enable you to form your own judgment upon the subject.

The stones immediately over the entrance are joggled into each other, the key-stone having a joggle on either side.—I have not observed this peculiarity in any other specimen of Norman masonry.—Between these porches apartments, along the interior of which runs a cornice, supported by grotesque corbels, and under it a row of windows, now principally blocked up, disposed in triplets, a trefoil-headed window being placed between two that are semi-circular, as seen in the accompanying drawing. The date of the origin of the trefoil-headed arch has been much disputed: these perhaps are some of the earliest, and they are unquestionably coeval with the building.

The stupid and disgraceful barbarism, which is now employing itself in the ruins of Jumieges, has long since annihilated the invaluable monuments which it contained.—In the Lady-Chapel of the conventual church was buried the heart of the celebrated Agnes Sorel, mistress of Charles VIIth, who died at Mesnil, about a league from this abbey, during the time when her royal lover was residing here.—Her death was generally attributed to poison; nor did the people hesitate in whispering that the fatal potion was administered by order of the Queen. Her son, the profligate tyrant Louis XIth, detested his father's concubine; and once, forgetting his dignity and his manhood, he struck the Dame de Beauté.—The statue placed upon the mausoleum represented Agnes kneeling and offering her heart to the virgin; but this effigy had been removed before the late troubles: a heart of white marble, which was at the foot of the tomb, had also disappeared. According to the annals of the abbey, they were destroyed by the Huguenots. The tomb itself, with various brasses inlaid upon it, remained undisturbed till the period of the revolution, when the whole memorial was removed, and even her remains were not suffered to rest in peace. The slab of black marble which covered them, and which bore upon its edges the French inscription to her memory, is still in existence; though it has changed its place and destination. The barbarians who pillaged the convent sold it with the rest of the plunder; and it now serves as a threshold to a house near the Mont aux Malades, at Rouen[17]. The inscription, which is cut in very elegant Gothic characters, is as follows: a part of it is, however, at present hidden by its position:—"Cy gist Agnes Surelle, noble damoiselle, en son vivant Dame de Roqueferriere, de Beaulté, d'Yssouldun, et de Vernon sur Seine, piteuse entre toutes gens, qui de ses biens donnoit largement aux gens d'église et aux pauvres; qui trespassa le neuvieme jour de Fevrier, l'an de grace 1449.—Priez Dieu pour elle."—It is justly to be regretted, that some pains are not taken for the preservation of this relic, which even now would be an ornament to the cathedral.—The manor-house at Mesnil, where the fair lady died, still retains its chimneys of the fifteenth century; and ancient paintings are discernible on the walls.

The monument in the church of St. Peter, generally known by the name of le tombeau des énervez, was of still greater singularity. It was an altar-tomb, raised about two feet above the pavement; and on the slabs were carved whole-length figures, in alto-relievo, of two boys, each about sixteen years of age, in rich attire, and ornamented with diadems, broaches, and girdles, all copiously studded with precious stones. Various traditions concerning this monument are recorded by authors, and particularly at great length by Father du Plessis[18].—The nameless princes, for such the splendor of their garb denotes them to have been, were considered, according to a tradition which prevailed from very early times, as the sons of Clovis and Bathilda, who, in the absence of their father, were guilty of revolt, and were punished by being hamstrung; for this is the meaning of the word énervez.—According to this tradition, the monks, in the thirteenth century, caused the monument to be ornamented with golden fleurs-de-lys, and added the following epitaph:—

"Hic in honore Dei requiescit stirps Clodovei,

Patris bellica gens, bella salutis agens.

Ad votum matris Bathildis poenituere,

Scelere pro proprio, proque labore patris."—

Three other lines, preserved by Yepez, in his chronicle, refer to the same tale, but accuse the princes of a crime of deeper die than mere rebellion against parental authority:—

"Conjugis est ultus probrum; nam in vincula tradit

Crudeles natos, pius impietate, simulque

Et duras pater, o Clodovee, piusque maritus."

Mabillon supposed the tomb to have been erected for Tassilo and his son; but I do not know how this conjecture is to be reconciled to the appearance of the statues, both representing persons of equal age. An examination of the grave at the time of the destruction of the abbey, might have afforded some interesting results; though, had any discovery been made, it would have been but a poor reward for the desolation which facilitated the research.

Footnotes:

[10] Immediately on the opposite side of the Seine, are extensive turf-bogs, which are of rare occurrence in this part of France; and in them grows the Andromeda polifolia, a plant that seems hitherto to have been discovered no where else in the kingdom.

[11] The following particulars relative to the territory of Jumieges, as well as the church, are curious: they are copied from an extract from the Life of St. Philibert, as given in the Neustria Pia, p. 262.—"Congruè sanè locus ille Gemmeticus est dictus, quippe qui instar gemmarum multivario sit decore conspicuus. Videas illic arborum comas sylvestrium, multigenos arborum fructus, solum fertile, prata virentia, hortorum flores suaveolentes, bortis gravidas vîtes, humum undique cinctam aquis, pascua pecorum uberrima, loca venationi apta, avium cantu circumsonantia. Sequana fluvius illic cernitur late ambiens: et deindè suo pergeus cursu, uno duntaxat commeantibus aditu relicto. Ibi mare increscens nunc eructat: nunc in sinum suum revolutum, navium

fert compendia, commercia plurimorum. Nihil illic deest; quicquid vehiculis pedestribus, et equestribus plaustris, et ratibus subministratur, abunde suppetit. Illic castrum condidere antiqui; ibi stant, in acie, illustria castra Dei: ibi præ desiderio paradisi suspirantes gemunt, quibus postea opus non erit, in flammis ultricibus, nihil profuturos edere gemitus. Ibi denique almus sacerdos, Philibertus, multiplici est laude et prædicatione efferendus: qui instar Patriarchæ Jacob, in animabus septuaginta, demigravit in hanc eremum, addito grege septemplici, propter septiformem gratiam spiritus sancti. Ibi enim eius prudentia construxit mœnia quadrata, turrita mole surgentia; claustra excipiendis adventantibus mirè opportuna. In his domus alma fulget; habitatoribus digna. Ab Euro surgit Ecclesia, crucis effigie, cujus verticem obtinet Beatissima Virgo Maria; Altare est ante faciem lectuli, cum Dente sanctiss, patris Philiberti, pictum gemmarum luminibus, auro argentoque comptum: ab utroque latere, Joannis et Columbani Aræ dant gloriam Deo; adherent verò a Boreâ, Dyonisii Martyris, et Germani Confessoris, ædiculæ; in dextrâ domus parte, sacellum nobile extat S. Petri; a latere habens S. Martini oratorium. Ad Austrum est S. Viri cellula, et petris habens margines; saxis cinguntur claustra camerata: is decor cunctorum animos oblectans, eum inundantibus aquis, geminus vergit ad Austrum. Habet autem ipsa domus in longum pedes ducentos nonaginta, in latum quinquaginta: singulis legere volentibus lucem transmittunt fenestræ vitreæ: subtus habet geminas ædes, alteras condendis vinis, alteras cibis apparandis accommodatas."

[12] Allusions to the cultivation of the vine at Jumieges, as then commonly practised, may be found in many other public documents of the fifteenth century: but we may come yet nearer our own time; for we know that, in the year 1500, there was still a vineyard in the hamlet of Conihoult, a dependence upon Jumieges, and that the wine called vin de Conihoult, is expressly mentioned among the articles of which the charitable donations of the monastery consisted.—We are told, too, that at least eighteen or twenty acres, belonging to the grounds of the abbey itself, were used as a vineyard as late as 1561.—At present, I believe,

vines are scarcely any where to be seen in Normandy, much north of Gaillon.

[13] In a charter belonging to the monastery, granted by Henry IInd, in 1159, (see Neustria Pia, p. 323) he gives the convent, "integritatem aquæ ex parte terræ Monachorum, et Graspais, si fortè capiatur."—The word Graspais is explained by Ducange to be a corruption of crassus piscis. Noel (in his Essais sur le Département de la Seine Inférieure, II, p. 168) supposes that it refers particularly to porpoises, which he says are still found in such abundance in the Seine, nearer its mouth, that the river sometimes appears quite black with them.

[14] The following account of the destruction of the monastery is extracted from William of Jumieges. (See Duchesne's Scriptores Normanni, p. 219)—"Dehinc Sequanica ora aggrediuntur, et apud Gemmeticum classica statione obsidionein componunt.... In quo quamplurima multitudo Episcoporum, seu Clericorum, vel nobilium laïcorum, spretis secularibus pompis, collecta, Christo Regi militatura, propria colla saluberrimo iugo subegit. Cuius loci Monachi, sive incolæ, Paganorum adventum comperientes, fugâ lapsi quædam suarum rerum sub terra occulentes, quædam secum asportantes, Deo juvante evaserunt. Pagani locum vacuum reperientes, Monasterium sanctæ Mariæ sanctíque Petri, et cuncta ædificia igne iniecto adurunt, in solitudinem omnia redigentes. Hac itaque patrata eversione, locus, qui tauto honoris splendore diu viguerat, exturbatis omnibus ac subuersis domibus, cœpit esse cubile ferarum et volucrum: maceriis in sua soliditate in sublime porrectis, arbustisque densissimis; et arborum virgultis per triginta fermè annorum curricula ubique a terra productis."

[15] The following are the proportions of the building, in French feet:—

Length of the church	265
Ditto of the nave	134
Width of ditto	62
Length of choir	43½
Width of ditto	31
Length of Lady-Chapel	63

Width of ditto	27
Height of central tower	124
Ditto of western towers	150

[16] Mr. Cotman has figured this porch, (Architectural Antiquities of Normandy, t. 4) but has, by mistake, called it "An Arch on the West Front of the Abbey Church."

[17] See a paper by M. Le Prevost in the Précis Analitique des Travaux de l'Académie de Rouen, 1815, p. 131.

[18] Histoire de la Haute Normandie, II, p. 260.

LETTER XVI
GOURNAY—CASTLE OF NEUFMARCHÉ—CASTLE AND CHURCH OF GISORS

(*Gisors, July*, 1818)

We are now approaching the western frontiers.—Gournay, Gisors, and Andelys, the objects of our present excursion, are disposed nearly in a line between the capitals of France and Normandy; and whenever war broke out between the two states, they experienced all the glory, and all the afflictions of warfare. This district was in fact a kind of debatable land; and hence arose the numerous strong holds, by which the country was once defended, and whose ruins now adorn the landscape.

The tract known by modern topographers, under the names of the *arrondissemens* of Gournay and of Andelys, constituted one of the general divisions of ancient Normandy, the *Pays de Bray*. It was a tract celebrated beyond every other in France, and, from time immemorial, for the excellence of the products of its dairies. The butter of Bray is an indispensable requisite at every fashionable table at Paris; and the *fromage de Neufchâtel* is one of the only two French cheeses which are honored with a place in the bill of fare at Véry 's at Grignon 's, or at Beauvilliers '.

The females of the district frequently passed us on the road, carrying their milk and eggs to the provincial metropolis. Accustomed as we are to the Norman costume, we still thought that the many-colored attire and long lappetted cap, of the good wife, of Bray, in conjunction with her steed and its trappings, was a most picturesque addition to the surrounding scenery. The large pannier on either side of the saddle leaves little room for the lady, except on the hinder parts of the poor beast; and there she sits, perfectly free and dégagée, without either pillion or stirrup, showing no small portion of her leg, and occasionally waving a little whip, ornamented in the handle with tufts of red worsted.—We had scarcely quitted the suburbs of Rouen before we found ourselves in Darnétal, a place that has risen considerably in importance, since the revolution, from the activity of its numerous manufacturers. Its population is composed entirely of individuals of this description, to whose pursuits its situation upon the banks of the Robec and Aubette is peculiarly favorable: the greater part of the goods manufactured here are coarse cloths and flannels. Before the revolution, the town belonged to the family of Montmorenci.—The rest of the ride offered no object of interest. The road, like all the main post-roads, is certainly wide and straight; but the French seem to think that, if these two points are but obtained, all the rest may be regarded as matter of supererogation. Hence, very little attention is paid to the surface of the highways: even on those that are most frequented, it is thought enough to keep the centre, which is paved, in decent repair: the ruts by the side are frequently so deep as to be dangerous; and in most cases the cross roads are absolutely impassable to carriages of every description, except the common carts of the country.—There is nothing in which England has a more decided superiority over France than in the facility of communication between its different towns; and there is also nothing which more decidedly marks a superiority of civilization. English travellers, who usually roll on the beaten track to and from the capital, return home full of praises of the French roads; but were they to attempt excursions among the country-towns and villages, their opinion would be wofully altered.—The forest of Feuillée extends about four leagues on each side of the road, between Rouen and Gournay. It adds little to the pleasantness of the ride: the trees are planted with regularity, and the side-branches are trimmed away almost to the very tops. Those therefore who expect overhanging branches, or the green-wood shade, in a French forest, will be sadly disappointed. On the contrary, when the wind blows across the road, and the sun shines down it, such a forest only adds to the heat and closeness of the way.

The country around Gournay is characterized by fertility and abundance; yet, in early times, the rich valley in which it is situated, was a dreary morass, which separated the Caletes from the Bellovacences. A causeway crossed the marshes, and formed the only road of communication between these tribes; and Gournay arose as an intermediate station. Therefore, even prior to the Norman æra, the town was, from its situation, a strong hold of note; and under the Norman dukes, Gournay necessarily became of still greater consequence, as the principal fortress on the French frontier; but the annexation of the duchy to the crown of France, destroyed this unlucky pre-eminence; and, at present, it is only known as a great staple mart for cheese and butter. Nor is it advantageously situated for trade; as there is no navigable river or means of water-carriage in its vicinity. The inhabitants therefore look forward with some anxiety to the completion of the projected canal from Dieppe.

Gournay is a small, clean, and airy place. The last two circumstances are no trifling recommendation to those who have just escaped from the dirt and closeness of Rouen. Its streets are completely those of a country town: the intermixture of wood and clay in the houses gives them a mean aspect, and there are scarcely two to be found alike, either in size, shape, color, or materials.—The records of Gournay begin in the reign of Rollo. That prince gave the town, together with the Norman portion of the Pays de Bray, to Eudes[19], a nobleman of his own nation, to be held as a fief of the duchy, under the usual military tenure. In one of the earliest rolls of Norman chieftains[20], the Lord of Gournay is bound, in case of war, to supply the duke with twelve soldiers from among his vassals, and to arm his dependants for the defence of his portion of the marches. Hugh, the son of Eudes de Gournay, erected a castle in the vicinity of the church of St. Hildebert, and the whole town was surrounded with a triple wall and double fosse. The place was inaccessible to an invading enemy, when these fosses were filled with the waters of the Epte; but Philip Augustus caused the protecting element to become his most powerful auxiliary. Willelmus Brito relates this siege with minuteness in his Philippiad, an heroic poem, devoted to the acts and deeds of the French monarch.—After advancing through Lions and Mortemer, Philip encamped before Gournay, thus described by the historical bard;—

"Non procul hinc vicum populosâ genta superbum,
Divitiis plenum variis, famâque celebrem,
Rure situm piano, munitum triplice muro,
Deliciosa nimis speciosaque vallis habebat.
Nomine GORNACUM, situ inexpugnabilis ipso,
Etsi nullus ei defensor ab intus adesset;
Cui multisque aliis præerat Gornacius HUGO.
Fossæ cujus erant amplæ nimis atque profundae
Quas sic Epta suo repleret flumine, posset
Nullus ut ad muros per eas accessus haberi.
Arte tamen sibi REX tali pessundedit ipsum.
Haud procul a muris stagnum pergrande tumebat,
Cujus aquam, pelagi stagnantis more, refusam
Urget stare lacu sinuoso terreus agger,
Quadris compactus saxis et cespite multo.
Hunc REX obrumpi medium facit, effluit inde
Diluvium immensum, subitâque voragine tota
Vallis abit maris in speciem, ruit impete vasto
Eluvies damnosa satis, damnosa colonis.

Municipes fugiunt ne submergantur, et omnis
Se populus villâ viduat, vacuamque relinquit.

Armis villa potens, muris munita virisque,
Arte capi nullâ metuens aut viribus ullis,
Diluvio capitur inopino....

REX ubi GORNACUM sic in sua jura redegit,
Indigenas omnes revocans ad propria, pacem
Indicit populis libertatemque priorem;
Deinde re-ædificat muros....

In 1350, after the death of Philip of Valois, Gournay was again separated from France, and given as a dower to Blanche of Navarre, the widow of that prince, who held it forty-eight years, when, after her death, it reverted to the crown. At the commencement of the following century, the town fell, with the rest of the kingdom, into the possession of the English; and once more, upon the demise of our sovereign, Henry Vth, formed part of the dower of the widowed queen. On her decease, it devolved upon her son; but a period of eleven years had scarcely elapsed, when the laws of conquest united it for a third time to the crown of France, in 1449.—From that period to the revolution, it was constantly in the possession of different noble families of the kingdom.

The name of Hugo de Gournay is enrolled amongst those who followed the conqueror into England, and who held lands in capite from him in this country[21]. Hugo was a man of eminent valor, and his services were requited by the grant of many large possessions; but, after all his military actions, he sought repose in the abbey of Bec, which had been enriched by his piety. His son, Girald, who married the sister of William, Earl Warren, accompanied Robert, Duke of Normandy, into the Holy Land; and the grandson of Girald was in the number of those who followed Richard Coeur-de-Lion in a similar expedition, and was appointed his commissioner, to receive the English share of the spoil, after the capture of Acre. He was also among the barons who rose against King John. Their descendants settled in very early times in our own county, where their possessions were extensive and valuable.

It was in Gournay that the unfortunate Arthur, heir to the throne of England, received the order of knighthood, together with the earldoms of

Brittany, Poitou, and Angers, from Philip Augustus, immediately previously to entering upon the expedition, which ultimately ended with his death; and, according to tradition, it was on this occasion that the town adopted for its arms the sable shield, charged with a knight in armor, argent[22].

Gournay has now no other remains of antiquity, except the collegiate church of St. Hildebert[23], which was founded towards the conclusion of the eleventh century, though it was scarcely completed at the end of the thirteenth. Hence the discrepancy of style observable in the architecture of its different parts. The west front, in which the windows are all pointed, was probably one of the last portions completed. The interior is principally of semi-circular architecture, with piers unusually massy, and capitals no less fanciful and extraordinary than those already noticed at St. Georges. Here, however, we have fewer monsters. The ornaments consist chiefly of foliage, and wreaths, and knots, and chequered work, and imitations of members of the antique capital. Some of the pillars, instead of ending in regular capitals, are surmounted by a narrow projecting rim, carved with undulating lines. It has been supposed that this ornament, which is quite peculiar to the church of St. Hildebert, is a kind of hieroglyphical representation of water.—Perhaps, it is the chamber of Sagittarius; or, perhaps, it is a fess wavy, to which the same signification has been assigned by heralds.—If this interpretation be correct, the symbol is allusive to the ancient situation of the town, built in the midst of a marsh, intersected by two streams, the Epte and the St. Aubin.

While we were on the point of setting out from Gournay, we had the pleasure of meeting Mr. Cotman, who landed a few days since at Dieppe, and purposes remaining in Normandy, to complete a series of drawings which he began last year, towards the illustration of the architectural antiquities of the duchy. He has joined our party, and we are likely to have the advantage of his society for some little time.

The village of Neufmarché, about a league from Gournay, on the right bank of the Epte, still retains a small part of its castle, built by Henry Ist, to command the passage of the river, and to serve as a barrier against the incursions of the French. Its situation is good, upon an artificial hill, surrounded by a fosse; and the principal entrance is still tolerably entire. But the rest is merely a shapeless heap of ruins: the interior is wholly under the plough; and the fragments of denudated walls preserve small remains of the coating of large square stones, which formerly embellished and protected them. Neufmarché, in the days of Norman sovereignty, was one of the strong holds of the duchy. The chroniclers[24] speak of the village as being defended by a fortress, in the reign of William the Conqueror. The church, too, with its semi-circular architecture, attests the antiquity of the station.

Long before we reached Gisors, we had a view of the keep of the castle, rising majestically above the town, which is indeed at present "une assez maussade petite ville, qui n'a guère qu'une rue." From its position and general outline, the castle, at first view, resembles the remains of Launceston, in Cornwall. It recalled to my mind the impressions of surprise, mixed with something approaching to awe, which seized me, when the first object that met my eyes in the morning (for it was late and dark when I reached Launceston) was the noble keep, towering immediately above my chamber windows, and so near, that it appeared as if I had only to open them and step into it. I do not mean to draw a parallel between the castles of Launceston and Gisors, and still less am I about to inquire into the relationship between the Norman and the Cornish fortresses. The lapse of twenty years has materially weakened my recollection of the latter, nor would this be a seasonable opportunity for such a disquisition: but the subject deserves investigation, the result of which may tend to establish the common origin of both, and to dissipate the day-dreams of Borlase, who longed to dignify the castellated ruins of the Cornish peninsula, by ascribing them to the Roman conquerors of Britain.

Gisors itself existed before the tenth century; but its chief celebrity was due to William Rufus, who, anxious to strengthen his frontiers against the power of the kings of France, caused Robert of Bellême to erect this castle, in 1097. Thus then we have a certain date; and there is no reason to believe, but that the whole of what is left us is really of the same æra, or of the following reign, in which it is known that the works were greatly augmented; for Henry Ist was completely a castle-builder. He was a prince who spared no pains in strengthening and defending the natural frontiers of his province, as the fortresses of Verneuil, Tillières, Nonancourt, Anet, Ivry, Château-sur-Epte, Gisors, and many others, abundantly testify. All these were either actually built, or materially strengthened by him.—This at Gisors, important from its strength and from its situation, was the source of frequent dissentions between the sovereigns of England and France, as well as the frequent witness of their plighted faith, and the scene of their festivities.—In 1119, a well-known interview took place here, between Henry Ist and Pope Calixtus IInd, who had travelled to France for the purpose of healing the schisms in the church, and who, after having accomplished that task, was desirous not to quit the kingdom till he had completed the work of pacification, by reconciling Henry to Louis le Gros, and to his brother, Robert. The speech of our sovereign upon this occasion, as recorded by Ordericus Vitalis[25], is a valuable document to the English historian: it sets forth, at considerable length, his various causes of grievance, whether real, imaginary, or invented, against the legal heir to our throne.—After a lapse of thirty-nine years, Louis le Jeune succeeded in annexing Gisors to the crown of France; but he resigned it to our Henry IInd, only three years subsequently, as a part of the marriage portion of his daughter, Margaret. It then remained with our countrymen till the conquest of the duchy by Philip Augustus; previously to which event, that sovereign and Henry met, in the year 1188, under an elm near Gisors, on the road to Trie, upon receiving the news of the capture of Jerusalem by the Sultan Saladin[26]. The monarchs, actuated by religious zeal, took up the cross, and mutually pledged themselves to suspend for a while their respective differences, and direct their united efforts against the common foe of the christian faith, Legends also tell that, during the conference, a miraculous cross appeared in the air, as if in ratification of the compact; and hence the inhabitants derive the armoria bearing of the town; gules, a cross engrailed or[27]. In 1197, Philip embellished Gisors with new buildings; and he retired hither the following year, after the battle of Courcelles, a conflict, which began by his endeavor to surprise Richard Coeur-de-Lion, but which ended with his total defeat. He had well nigh lost his life during the flight, by his horse plunging with him, all armed as he was, into the Epte.—He took refuge in Gisors; and the golden gate of the town commemorated his gratitude. With eastern magnificence, he caused the entire portal to be covered with gold; and the statue of the Virgin, which surmounted it, received the same splendor.

During the wars between France and England, in the fifteenth century, Gisors was repeatedly won and lost by the contending parties. In later and more peaceable times, it has been only known as the provincial capital of the bailiwick of Gisors, and of the Norman portion of the Vexin.

The castle consists of a double ballium, the inner occupying the top of a high artificial mound, in whose centre stands the keep. The whole of the fortress is of the most solid masonry. Previously to the discovery of cannon, it could scarcely be regarded otherwise than as impregnable, for the site which it occupies is admirably adapted for defence; and the walls were as strong as art could make them.—The outer walls were of great extent: they were defended by two covered ways, and flanked by several towers, of various shapes.—In the inclosed sketch, you will observe a circular tower, which is perhaps more perfect than any of the rest. The two entrances which led to the inner wards, were defended by more massy towers, strengthened with portcullises and draw-bridges.

Castle at Gisors

The conical mound is almost inaccessible, on account of its steepness. The summit is inclosed by a circular wall of considerable height, pierced with loop-holes, and strengthened at regular intervals with buttresses, most of which are small and shallow, and resemble such as are found in the Norman churches. Those, however, which flank the entrance of the keep, are of a different character: they project so boldly, that they may rather be considered as bastions or solid turrets.—The dungeon rises high above all the rest, a lofty octagon tower, with a turret on one side of the same shape, intended to receive the winding staircase, which still remains, but in so shattered a state, that we could not venture to ascend it. The shell of the keep itself is nearly perfect, and is also varied in its outline with projecting piers.—Within the inner ballium, we discovered the remains of the castle-chapel. More than half, indeed, of the building is destroyed, but the east end is standing, and is tolerably entire. The roof is vaulted and groined: the groins spring from short pillars, whose capitals are beautifully sculptured with foliage; The architecture of the whole is semi-circular; but I should apprehend it to be posterior to any part of the fortress.—The inside of the castle serves at this time for a market-hall: the fosse, now dry and planted with trees, forms a delightful walk round the whole.

We were much disappointed by the church of Gisors; in the illustration of the details of which, Millin is very diffuse. The building is of considerable magnitude; its proportions are not unpleasing, and it contains much elaborate sculpture; but the labor has been ill bestowed, having been lavished without any attention to consistency. It is throughout a jumble of Roman and Gothic, except that the exterior of the north transept is wholly Gothic. Some of the little figures which decorate it are very gracefully carved, especially in the drapery. A pillar in the south aisle, entwined by spiral fillets, is of great singularity and beauty. The dolphin is introduced in each pannel, and the heraldic form of this fish harmonizes with the gentle curve of the field upon which it is sculptured. A crown of fleurs-de-lys surrounds the columns at mid-height. These symbols, as I believe I observed on a former occasion, are often employed as ornaments by the French architects. The church, which is dedicated to the twin saints, St. Gervais and St. Protais, is the work of different æras, but principally of the latter half of the sixteenth century, a time when, as a Frenchman told me, "l'on commença à bâtir dans le beau style Romain."—The man who made the observation was of the lower order of society, one of the swinish multitude, who, in England, never dream about styles in architecture. I mention the circumstance, for the sake of pointing out the difference that exists in these matters between the two countries.

Here, every man, gentle or simple, educated or uneducated, thinks himself qualified and bound to deliver his opinion on objects connected with the fine arts; and though such opinions are of necessity commonly crude, and sometimes absurd, they, on the other hand, frequently display a degree of feeling, and occasionally of knowledge, that surprises you. It may be true indeed, as Dr. Johnson said, with some illiberality, of our brethren across the Tweed, that though "every man may have a mouthful, no one has a belly full;" but it still marks a degree of national refinement, that any attention whatever is bestowed upon such subjects. This smattering of knowledge, accompanied with the constant readiness to communicate it, is also agreeable to a stranger. Except in a few instances at Rouen, I never failed to find civility and attention among the French. To the ladies of our nation they are uniformly polite though occasionally their compliments may appear of somewhat a questionable complexion; as it happened to a female friend of mine to be told, while drawing the church of St, Ouen, "qu'elle avait de l'esprit comme quatre diables."

Footnotes:

[19] Histoire de la Haute Normandie, I, p. 18.

[20] Duchesne, Scriptores Normanni, p. 1046.

[21] Duchesne, Scriptores Normanni, p. 1129.

[22] Histoire de la Haute Normandie, I. p. 20.

[23] See Cotman's Architectural Antiquities of Normandy, plates 38-41.

[24] Ordericus Vitalis, in Duchesne's Scriptores Normanni, p. 490, 491, 606.

[25] Duchesne, Scriptores Normanni, p. 865.

[26] Some writers say that the real cause of their meeting was to settle a difference of long standing.—Hoveden, as quoted in the Concilia Normannica, I. p. 92, tells us, that Henry was upon the point of sailing for England, when tidings were brought him that Philip had collected a great force, with which he threatened to lay Normandy waste, unless the British monarch surrendered to him Gisors with its dependencies, or caused his son Richard, Count of Poitou, to marry Alice, sister of the French king;—"Quod cùm regi Angliæ constaret, reversus est in Normanniam; et, accepte colloquio inter ipsum et Regem Franciæ inter Gisortium et Trie, XII. Kalendas Februarii, die S. Agnetis V. et Martyris, convenerunt illuc cum Archiepiscopis, et Episcopis et Comitibus, et Baronibus regnoram suorum. Cui colloquio interfuit Archiepiscopus Tyri, qui repletus spiritu sapientiæ et intellectus, miro modo prædicavit verbum Domini coram regibus et principibus. Et convertit corda eorum ad crucem capiendam; et qui priùs hostes erant, illo prædicante, et Deo co-operante, facti sunt amici in illa die, et de manu ejus crucem receperunt: et in eadem hora apparuit super eos signum crucis in cœlo. Quo viso miraculo, plures catervatim ruebant ad susceptionem crucis. Prædicti verò reges in susceptionem crucis, ad cognoscendum gentem suam, signum sibi et suis providerunt. Rex namque Franciæ et gens sua receperunt cruces rubeas et Rex Angliæ cum gente sua suscepit cruces virides: et sic unusqnisque ad providendum sibi et itineri suo necessaria, reversus est in regionem suam."

[27] In 1555, an addition was made to this coat of a chief azure, charged with three fleurs-de-lys, or, by the command of Henry IInd of France, to commemorate his public entry into Gisors.

LETTER XVII ANDELYS—FOUNTAIN OF SAINT CLOTILDA—LA GRANDE MAISON—CHÂTEAU GAILLARD—ECOUIS

(*Ecouis, July*, 1818)

Our evening journey from Gisors to Andelys, was not without its inconveniences.—The road, if road it may be called, was sometimes merely a narrow ravine or trench, so closely bordered by trees and underwood, that our vehicle could scarcely force its way; and sometimes our jaded horses labored along a waggon-way which wound amidst an expanse of corn-fields. Our postilion had earnestly requested us to postpone our departure till the following morning; and he swore and cursed most valiantly during the whole of his ride. On our arrival, however, at Andelys, a few kind words from my companions served to mitigate his ire; and as their eloquence may have been assisted by a few extra sous, presented to him at the same time, his nut-brown countenance brightened up, and all was tranquillity.

Andelys is a town, whose antiquity is not to be questioned: it had existence in the time of the venerable Bede, by whom it is expressly mentioned, under its Latin appellation, Andilegum[28]. The derivation of this name has afforded employment to etymologists. The syllable and enters, as it is said, into the composition of the names of sundry places, reported to be founded by Franks, and Saxons, and Germans; and therefore it is agreed that a Teutonic origin must be assigned to Andelys. But, as to the import of this same syllable, they are all of them wholly at a loss.—The history of Andelys is brief and unimportant, considering its antiquity and situation. It was captured by Louis le Gros in the war which he undertook against Henry Ist, in favour of Clito, heir of the unfortunate Duke Robert; and his son, Louis le Jeune, in 1166, burned Andelys to the ground, thus revenging the outrages committed by the Anglo-Normans in France: in 1197, it was the subject of the exchange which I have already mentioned, between Richard Coeur-de-Lion and Walter, Archbishop of Rouen; and only a few years afterwards it passed by capitulation into the possession of Philip Augustus, when the murder of Arthur of Brittany afforded the French sovereign a plausible pretext for dispossessing our worthless monarch of his Norman territory.

What Andelys wants, however, in secular interest, it makes up in sanctity. Saint Clotilda founded a very celebrated monastery here, which was afterwards destroyed by the Normans.—If we now send our ripening daughters to France, to be schooled and accomplished, the practice prevailed equally amongst our Anglo-Saxon ancestors; and we learn from Bede, that Andelys was then one of the most fashionable establishments[29]. However, we must not forget that the fair Elfleda, and the rosy Ælfgiva, were so taugh in the convent, as to be fitted only for the embraces of a celestial husband—a mode of matrimony which has most fortunately become obsolete in ou days of increasing knowledge and civilization.

After the destruction of the monastery by the Normans, it was neve rebuilt; yet its sanctity is not wholly lost. At the behest of Clotilda, the waters of the fountain of Andelys were changed into wine for the relief of the weary labourer, and the tutelary saint is still worshipped by the faithful.

It was our good fortune to arrive at Andelys on the vigil of the festival of Saint Clotilda. The following morning, at early dawn, the tolling bell announced the returning holiday; and then we saw the procession advance, priests and acolytes bearing crosses and consecrated banners and burning tapers, followed by a joyous crowd of votaries and pilgrims. We had wished to approach the holy well; but the throng thickened around it, and we were forced to desist. We could not witness the rites, whatever they were, which were performed at the fountain; and long after they had concluded, it was still surrounded by groups of women, some idling and staring, some asking charity and whining, and some conducting their little ones to the salutary-fountain. Many are the infirmities and ailments which are relieved through the intercession of Saint Clotilda, after the patient has been plunged in the gelid spring. A Parisian sceptic might incline to ascribe a portion of their cures to cold-bathing and ablution; but, at Andelys, no one ever thought of diminishing the veneration, inspired by the Christian queen of the founder of the monarchy. Several children were pointed out to us, heretical strangers, as living proofs of the continuance of miracles in the Catholi church. They had been cured on the preceding anniversary; for it is only or Saint Clotilda's day that her benign influence is shed upon the spring.

Andelys possesses a valuable specimen of ancient domestic architecture. The Great House[30] is a most sumptuous mansion, evidently of the age of Francis Ist; but I could gain no account of its former occupants or history. I must again borrow from my friend's vocabulary, and say, that it is built in the "Burgundian style." In its general outline and character, it resembles the house in the Place de la Pucelle, at Rouen. Its walls, indeed, are not covered with the same profusion of sculpture; yet, perhaps, its simplicity is accompanied by greater elegance.—The windows are disposed in three divisions, formed by slender buttresses, which run up to the roof. They are square-headed, and divided by a mullion and transom.—The portal is in the centre: it is formed by a Tudor arch, enriched with deep mouldings, and surmounted by a lofty ogee, ending with a crocketed pinnacle, which transfixes the cornice immediately above, as well as the sill of the window, and then unites with the mullion of the latter.—The roof takes a very high pitch.—A figured cornice, upon which it rests, is boldly sculptured with foliage.—The chimneys are ornamented by angular buttresses.—All these portions of the building assimilate more or less to our Gothic architecture of the sixteenth century; but a most magnificent oriel window, which fills the whole of the space between the centre and left-hand divisions, is a specimen of pointed architecture in its best and purest style. The arches are lofty and acute. Each angle is formed by a double buttress, and the tabernacles affixed to these are filled with statues. The basement of the oriel, which projects from the flat wall of the house, after the fashion of a bartizan, is divided into compartments, studded with medallions, and intermixed with tracery of great variety and beauty. On either side of the bay, there are flying buttresses of elaborate sculpture, spreading along the wall.—As, comparatively speaking, good models of ancient domestic architecture are very rare, I would particularly recommend this at Andelys to the notice of every architect, whom chance may conduct to Normandy.—This building, like too many others of the same class in our own counties of Norfolk and Suffolk, is degraded from its station. The great house is used merely as a granary, though, by a very small expence, it might be put into habitable repair. The stone retains its clear and polished surface; and the massy timbers are undecayed.—The inside corresponds with the exterior, in decorations and grandeur: the chimney-pieces are large and elaborate, and there is abundance of sculpture on the ceilings and other parts which admit of ornament.

The French, in speaking of Andelys, commonly use the plural number, and say, *les Andelys*, there being a smaller town of the same name, within the distance of a mile: hence, the larger, all inconsiderable as it is, and though it scarcely contains two thousand inhabitants, is dignified by the appellation of *le Grand Andelys*.

As the French seldom neglect the memory of their eminent men, I was rather disappointed at not finding any tribute to the glory of Poussin, nor any object which could recal his name.—The great master of the French school was born at Andelys, in 1594, of poor but noble parents. The talents of the painter of the Deluge overcame all obstacles. Young Poussin, with barely a sufficiency to buy his daily bread, found means of making his abilities known in the metropolis to such advantage, as enabled him to proceed to Rome, where the patronage of the Cavaliere Marino smoothed his way to that splendid career, which terminated only with his life.—And yet I doubt if the example of Poussin has, on the whole, been favorable to the progress of French art. Horace Walpole, in his summary of the excellencies and defects of great painters, observed with much justice, that "Titian wanted to have seen the antique; Poussin to have seen Titian." The observation referred principally to the defective coloring, which is admitted to exist in the greater part of the works of the painter of Andelys. But Poussin, considered as a model for imitation, and especially as a model for the student, is liable to a more serious objection.—He was a total stranger to real nature:—classical taste, indeed, and knowledge, and grace, and beauty, pervade all his works; but it is a taste, and a knowledge, and a grace, and a beauty, formed solely upon the contemplation of the antique. Horace's adage, that "decipit exemplar vitiis imitabile," has been remarkably verified in the case of Poussin; and I am mistaken, if the example set by him, which has been rigorously followed in the French school, even down to the present day, has not contributed more than any thing else to that statuary style in forms, and that coldness in coloring, which every one, who is not born in France, regrets to see in the works of the best of their artists.—The learned Adrian Turnebus was also a native of Andelys; and the church is distinguished as the burial-place of Corneille.

I doubt, however, whether we should have travelled hither, had we not been attracted by the celebrity of the castle, called *Château Gaillard,* erected by Richard Coeur-de-Lion, in the immediate vicinity of Le Petit Andelys.—Our guide, a sturdy old dame, remonstrated strongly against our walking so far to look at a mere heap of stones, nothing comparable to the fine statue of Clotilda, of which, if we would but have a little patience, we might still procure a sight.—Our expectations respecting the castle were more than answered. Considered as to its dimensions and its situation, it is by far the finest castellated ruin I ever saw. Conway, indeed, has more beauty; but Château Gaillard is infinitely superior in dignity. Its ruins crown the summit of a lofty rock, abruptly rising from the very edge of the Seine, whose sinuous course here shapes the adjoining land into a narrow peninsula. The chalky cliffs on each side of the castle, are broken into hills of romantic shape, which add to the impressive wildness of the scene. The inclosed sketch will give you an idea, though a very faint one, of the general appearance of the castle at a distance. Towards the river, the steepness of the cliff renders the fortress unassailable: a double fosse of great depth, defended by a strong wall, originally afforded almost equal protection on the opposite side.

The circular keep is of extraordinary strength; and in its construction it differs wholly from any of our English dungeon-towers.—It may be described as a cylinder, placed upon a truncated cone. The massy perpendicular buttresses, which are ranged round the upper wall, from which they project considerably, lose themselves at their bases in the cone from which they arise. The building, therefore, appears to be divided into two stories. The wall of the second story is upwards of twelve feet in thickness. The base of the conical portion is perhaps twice as thick.—It seldom happens that the military buildings of the middle ages have such a talus or slope, on the exterior face, agreeing with the principles of modern fortification, and it is difficult to guess why the architect of Château Gaillard thought fit to vary from the established model of his age. The masonry is regular and good. The pointed windows are evidently insertions of a period long subsequent to the original erection.

The inner, ballium is surrounded by a high circular wall, which consists of an uninterrupted line of bastions, some semi-circular and others square.—The whole of this part of the castle remains nearly perfect. There are also traces of extensive foundations in various, directions, and of great out-works. Château Gaillard was in fact a citadel, supported by numerous smaller fortresses, all of them communicating with the strong central hold, and disposed so as to secure every defensible post in the neighborhood. The wall of the outer ballium, which was built of a compact white and grey stone, is in most places standing, though in ruins. The original facing only remains in those parts which are too elevated to admit of its being removed with ease.—Beneath the castle, the cliff is excavated into a series of subterraneous caverns, not intended for mere passages or vaults, as at Arques and in most other places, but forming spacious crypts, supported by pillars roughly hewn out of the living rock, and still retaining every mark of the workman's chisel.

It will afford some satisfaction to the antiquary to find, that the present appearance of the castle corresponds in every important particular with the description given by Willelmus Brito, who beheld it within a few years after its erection, and in all its pride. Every feature which he enumerates yet exists, unaltered and unobliterated:—

"Huic natura loco satis insuperabile per se
Munimeu dederat, tamen insuperabiliorem
Arte quidem multa Richardus fecerat illum.
Duplicibus muris extrema clausit, et altas
Circuitum docuit per totum surgere turres,
A se distantes spatiis altrinsecus æquis;
Eruderans utrumque latus, ne scandere quisquam
Ad muros possit, vel ab ima repere valle.
Hinc ex transverso medium per planitiei
Erigitur murus, multoque labore cavari
Cogitur ipse silex, fossaque patere profunda,
Faucibus et latis aperiri vallis ad instar;
Sic ut quam subito fiat munitio duplex
Quæ fuit una modo muro geminata sequestro.
Ut si forte pati partem contingeret istam
Altera municipes, queat, et se tuta tueri.
Inde rotundavit rupem, quæ celsior omni
Planitie summum se tollit in aera sursum;
Et muris sepsit, extremas desuper oras
Castigansque jugi scrupulosa cacumina, totum
Complanat medium, multæque capacia turbæ
Plurima cum domibus habitacula fabricat intus.
Umboni parcens soli, quo condidit arcem.
Hic situs iste decor, munitio talis honorem
Gaillardæ rupis per totum prædicat orbem."

The keep cannot be ascended without difficulty. We ventured to scale it; and we were fully repaid for our labor by the prospect which we gained. The Seine, full of green willowy islands, flows beneath the rock in large lazy windings: the peninsula below is flat, fertile, and well wooded: on the opposite shores, the fantastic chalky cliffs rise boldly, crowned with dark forests.

I have already once had occasion to allude to the memorable strife occasioned by the erection of Château Gaillard, which its royal founder is reported to have so named by way of mockery. In possession of this fortress, it seemed that he might laugh to scorn the attacks of his feudal liege lord.—The date of the commencement of the building is supposed to have been about the year 1196, immediately subsequent to the treaty of Louviers, by which, Richard ceded to Philip Augustus the military line of the Epte, and nearly the whole of the Norman Vexin. By an express article of the treaty, neither party was allowed to repair the fortifications of Andelys; and Philip was in possession of Gisors, as well as of every other post that might have afforded security to the Normans. Thus the frontiers of the duchy became defenceless; but Richard, like other politicians, determined to evade the spirit of the treaty, adhering nevertheless to its letter, by the erection of this mighty bulwark.—The building arose with the activity of fear. Richard died in 1199, yet the castle must have been completely habitable in his life-time, for not a few of his charters are dated from Château Gaillard, which he terms "his beautiful castle of the rock."—Three years only had elapsed from the decease of this monarch, when Philip Augustus, after having reduced another castle, erected at the same time upon an island opposite the lesser Andelys, encamped before Château Gaillard, and commenced a siege, which from its length, its horrors, and the valor shewn on either side, has ever since been memorable in history.—Its details are given at great length by Father Daniel; and Du Moulin briefly enumerates a few of the stratagems to which the French King was obliged to have recourse; for, as the reverend author observes, "to have attempted to carry the place by force, would have been to have exposed the army to certain destruction; while to have tried to scale the walls, would have required the aid of Dædalus, with the certainty of a fall, as fatal as that of Icarus;" and without the poor consolation of

"…. vitreo daturus
Nomina ponto."—

The castle, commanded by Roger de Lacy, defied the utmost efforts of Philip for six successive months.—So great was its size; that more than two thousand two hundred persons, who did not form a part of the garrison, were known to quit the fortress in the course of the siege, compelled to throw themselves upon the mercy of the besiegers. But they found none; and the greater part of these unfortunate wretches, alternately suppliants to either host, perished from hunger, or from the weapons of the contending parties. At length the fortress yielded to a sudden assault. Of the warriors, to whose valor it had been entrusted, only thirty-six remained alive. John, ill requiting their fidelity, had already abandoned them to their fate.

Margaret of Burgundy, the queen of Louis Xth, and Blanche, the consort of his brother, Charles le Bel, were both immured in Château Gaillard, in 1314. The scandalous chronicle of those times will explain the causes of their imprisonment. Margaret was strangled by order of her husband. Blanche, after seven years' captivity, was transferred to the convent of Maubuisson, near Pontoise, where she continued a recluse till her death—In 1331, David Bruce, compelled to flee from the superior power of the third Edward, found an asylum in Château Gaillard; and here, for a time, maintained the pageantry of a court.—Twenty-four years subsequently, when Charles the Bad, king of Navarre, was sent as a captive from Rouen to Paris, he was confined here, during one night, by order of the dauphin, who had made him his prisoner by treachery, whilst partaking of a banquet.—In the following century Château Gaillard braved the victorious arms of Henry Vth; nor was it taken till after a siege of sixteen months. The garrison only consisted of one hundred and twenty men; yet this scanty troop would not have yielded, had not the ropes, by which they drew up their water-buckets[31], been worn out and destroyed.—During the same reign, it was again taken and lost by the French, into whose hands it finally fell in 1449, when Charles VIIth commanded the siege in person. Even then, however it stood a long siege; and it was almost the last of the strong-holds of Normandy, which held out for the successors of the ancient dukes. After the re-union of the duchy, it was not destroyed, or suffered to fall into decay, like the greater number of the Norman fortresses: during the religious wars, it still continued to be a formidable military post, as well as a royal palace; and it was honored by the residence of Henry IVth, whose father, Anthony of Bourbon, died here in 1562.—Its importance ceased in the following reign.—The inhabitants of the adjacent country requested the king to order that the castle should be dismantled. They dreaded, lest its towers should serve as an asylum to some of the numerous bands of marauders, by whom France was then infested. It was consequently undermined and reduced to its present state of ruin.

We did not again attempt to pay our devotions at the shrine of Saint Clotilda, and we found no interesting object in the church of Andelys which could detain us. We therefore proceeded without delay to Ecouis, where we were assured that the church would gratify our curiosity.—This building has an air of grandeur as it is seen rising above the flat country; and it is of a singular shape, the ground-plan being that of a Greek cross. The exterior is plain and offers nothing remarkable: the interior retains statues of various saints, which, though not very ancient or in very good taste, are still far from being inelegant. Saint Mary, the Egyptian, who is among them, covered with her tresses, which may easily be mistaken for a long plaited robe, is a saint of unfrequent occurrence in this part of France. In the choir are several tomb stones, with figures engraved upon them, their faces and hands being inlaid with white marble.—In this part of the building also remains the tomb of John Marigni, archbishop of Rouen, with his effigy of fine white marble, in perfect preservation. The face is marked with a strong expression of that determined character, which he unquestionably possessed. When he was sent as an ambassador to Edward IIIrd, in 1342, he made his appearance at the English court in the guise of a military man, and not as a minister of peace; and we may doubt whether his virtues qualified him for the mitre. If even a Pope, however, in latter days, commanded a sculptor to pourtray him with a sword in his hand, the martial tendency of an archbishop may well be pardoned in more turbulent times. The following distich, from his epitaph, alludes to his achievements:—

"Armis præcinctus, mentisque charactere cinctus,

Dux fuit in bellis, Anglis virtute rebellis."

The unfortunate Enguerrand de Marigni, brother of the archbishop, and lord treasurer under Philip the Fair, was the founder of this church. At the instigation of the king's uncle, Enguerrand was hanged without trial, and his family experienced the most bitter persecution. His body, which had at first been interred in the convent of the Chartreux, at Paris, was removed hither in 1324; and his descendants obtained permission, in 1475, to erect a mausoleum to his memory. But the king, at the same time that he acceded to their petition, added the express condition[32], that no allusion should be made to Marigni's tragical end. The monument was destroyed in the revolution; but the murder of the treasurer is one of those "damned spots," which will never be washed out of the history of France.—Charles de Valois soon felt the sting of remorse; and within a year from the wreaking of his vengeance, he caused alms to be publicly distributed in the streets of Paris, with an injunction to every one that received them, "to pray to God for the souls of Enguerrand de Marigni, and Charles de Valois, taking care to put the subject first[33]."—In the church at Ecouis, was formerly the following epitaph, whose obscurity has given rise to a variety of traditions:—

"Ci gist le fils, ci gist la mere,

Ci gist la soeur, ci gist le frère,

Ci gist la femme, et le mari;

Et ci ne sont que deux ici[34]."

Other inscriptions of the same nature are said to have existed in England. Goube[35] supposes that this one is the record of an incestuous connection; but we may doubt whether a less sinful solution may not be given to the enigma.

Footnotes:

[28] Andelys is also called in old deeds Andeleium and Andeliacum.

[29] "Seculo septimo, cum pauca essent in regione Anglorum monasteria, hunc morem in illâ gente fuisse, ut multi ex Britanniâ, monastiae conversationis gratiâ, Francorum monasteria adirent, sed et filias suas eisdem erudiendas ac sponso coelesti copulandas mitterent, maximè in Brigensi seu S. Farae monasterio, et in Calensi et in Andilegum monasterio."—Bede, Hist. lib. III. cap. 8.

[30] Cotman's Architectural Antiquities of Normandy, plate 15.—In a future portion of his work, Mr. Cotman designs devoting a second plate exclusively to the oriel in the east front of this building.

[31] Monstrelet, Johnes' Translation, II. p. 242.

[32] The letter of this stipulation appears to have been attended to much more than its spirit for at the top of the monument were five figures:—Our Savior seated in the centre, as if in the act of pronouncing sentence; on either side of him, an angel; and below, Charles de Valois and Enguerrand de Marigni; the former on the right of Christ, crowned with the ducal coronet; the other, on the opposite side, in the guise and posture of a suppliant, imploring the divine vengeance for his unjust fate.—Histoire de la Haute Normandie, II. p. 338.

[33] Montfaucon, Monumens de la Monarchie Française, II. p. 220.

[34] In a collection of epitaphs printed at Cologne, 1623, under the title of Epitaphia Joco-seria, I find the same monumental inscription, with the observation, that it is at Tournay, and with the following explanation.—"De pari conjugum, posteà ad religionem transeuntium et in eâ præfectorum. Alter fuit Franciscanus; altera verò Clarissa."

[35] Histoire du Duché de Normandie, III. p. 15.

LETTER XVIII
EVREUX—CATHEDRAL—ABBEY OF ST. TAURINUS—ANCIENT HISTORY

(*Evreux, July,* 1818.)

Our journey to this city has not afforded the gratification which we anticipated.—You may recollect Ducarel's eulogium upon the cathedral, that it is one of the finest structures of the kind in France.—It is our fate to be continually at variance with the doctor, till I am half inclined to fear you may be led to suspect that jealousy has something to do with the matter, and that I fall under the ban of the old Greek proverb,—

"Και κεραμευς κεραμει Φθονεει και τεκτονι τεκτων."—

[English. Not in Original: The potter is jealous of the potter, as the builder is jealous of the builder.]

As for myself, however, I do hope and trust that I am marvellously free from antiquarian spite.—And in this instance, our expectations were also raised by the antiquity and sanctity of the cathedral, which was entirely rebuilt by Henry Ist, who made a considerate bargain with Bishop Audinus[36], by which he was allowed to burn the city and its rebellious inhabitants, upon condition of bestowing his treasures for the re-construction of the monasteries, after the impending conflagration. The church, thus raised, is said by William of Jumieges[37], to have surpassed every other in Neustria; but it is certain that only a very small portion of the original building now remains. A second destruction awaited it. Philip Augustus, who desolated the county of Evreux with fire and sword, stormed the capital, sparing neither age nor sex; and all its buildings, whether sacred or profane, were burnt to the ground. Hoveden, his friend, and Brito, his enemy, both bear witness to this fact—the latter in the following lines:—

"... irarum stimulis agitatus, ad omne
Excidium partis adversæ totus inardens,
Ebroicas primò sic incineravit, ut omnes
Cum domibus simul ecclesias consumpserit ignis."—

The church, in its present state, is a medley of many different styles and ages: the nave alone retains vestiges of early architecture, in its massy piers and semi-circular arches: these are evidently of Norman workmanship, and are probably part of the church erected by Henry.—All the rest is comparatively modern.—The western front is of a debased Palladian style, singularly ill adapted to a Gothic cathedral. It is flanked with two towers, one of which ends in a cupola, the other in a short cone.—The central tower, which is comparatively plain and surmounted by a high spire, was built about the middle of the fifteenth century, during the bishopric of the celebrated John de Balue, who was in high favor with Louis XIth, and obtained from that monarch great assistance towards repairing, enlarging, and beautifying his church. The roof, the transept towards the palace, the sacristy, the library, and a portion of the cloisters, are all said to have been erected by him[38].—The northern transept is the only part that can now lay claim to beauty or uniformity in its architecture: it is of late and bastard Gothic; yet the portal is not destitute of merit: it is evidently copied from the western portal of the cathedral at Rouen, though far inferior in every respect, and with a decided tendency towards the Italian style. Almost every part of it still appears full of elaborate ornaments, though all the saints and bishops have fled from the arched door-way, and the bas-relief which was over the entrance has equally disappeared.

Ducarel[39] notices four statues of canons, attached to a couple of pillars at the back of the chancel.—We were desirous of seeing authentic specimens of sculpture of a period at least as remote as the conquest; and, as the garden belonging to the prefect, the Comte de Goyon, incloses this portion of the church, we requested to be allowed to enter his grounds. Leave was most obligingly granted, and we received every attention from the prefect and his lady; but we could find no traces of the objects of our search. They were probably destroyed during the revolution; at which time, the count told us that the statues at the north portal were also broken to pieces. At Evreux, the democrats had full scope for the exercise of their iconoclastic fury. Little or no previous injury had been done by the Calvinists, who appear to have been unable to gain any ascendency in this town or diocese, at the same time that they lorded it over the rest of Normandy. Evreux had been fortified against heresy, by the piety and good sense of two of her bishops: they foresaw the coming storm, and they took steps to redress the grievances which were objects of complaint, as well as to reform the church-establishment, and to revise the breviary and the mass-book.—Conduct like this seldom fails in its effect; and the tranquil by-stander may regret that it is not more frequently adopted by contending parties.

The interior of the cathedral is handsome, though not peculiar. Some good specimens of painted glass remain in the windows; and, in various parts of the church, there are elegant tabernacles and detached pieces of sculpture, as well in stone as in wood. The pulpit, in particular, is deserving of this praise: it is supported on cherubs' heads, and is well designed and executed.

The building is dedicated to the Virgin: it claims for its first bishop, Taurinus, a saint of the third century, memorable in legendary tale for a desperate battle which he fought against the devil. Satan was sadly drubbed and the bishop wrenched off one of his horns[40]. The trophy was deposited in the crypt of his church, where it long remained, to amuse the curious, and stand the nurses of Evreux in good stead, as the means of quieting noisy children.—The learned Cardinal Du Perron succeeded to St. Taurinus, though at an immense distance of time. He was appointed by Henry IVth, towards whose conversion he appears to have been greatly instrumental, as he was afterwards the principal mediator, by whose intercession the Pope was induced to grant absolution to the monarch. The task was one of some difficulty: for the court of Spain, then powerful at the Vatican, used all their efforts to prevent a reconciliation, with a view of fomenting the troubles in France.—Most of the bishops of this see appear to have possessed great piety and talent.

I have already mentioned to you, that the fraternity of the Conards was established at Evreux, as well as at Rouen. Another institution, of equal absurdity, was peculiar, I believe, to this cathedral[41]. It bore the name of the Feast of St. Vital, as it united with the anniversary of that saint, which is celebrated on the first of May: the origin of the custom may be derived from the heathen Floralia, a ceremony begun in innocence, continued to abomination. At its first institution, the feast of St. Vital was a simple and a natural rite: the statues of the saints were crowned with garlands of foliage, perhaps as an offering of the first-fruits of the opening year. In process of time, branches were substituted for leaves, and they were cut from the growing trees, by a lengthened train of rabble pilgrims.—The clergy themselves headed the mob, who committed such devastation in the neighboring woods, that the owners of them were glad to compromise for the safety of their timber, by stationing persons to supply the physical, as well as the religious, wants of the populace. The excesses consequent upon such a practice may easily be imagined: the duration of the feast was gradually extended to ten days; and, during this time, licentiousness of all kinds prevailed under the plea of religion. To use the words of a manuscript, preserved in the archives of the cathedral, they played at skittles on the roof of the church, and the bells were kept continually ringing. These orgies, at length, were quelled; but not till two prebendaries belonging to the chapter, had nearly lost their lives in the attempt.—Hitherto, indeed, the clergy had enjoyed the merriment full as well as the laity. One jolly canon, appropriately named Jean Bouteille, made a will, in which he declared himself the protector of the feast; and he directed that, on its anniversary, a pall should be spread in the midst of the church, with a gigantic bottle in its centre, and four smaller ones at the corners; and he took care to provide funds for the perpetuation of this rebus.

The cathedral offers few subjects for the pencil.—As a species of monument, of which we have no specimens in England, I add a sketch of a Gothic *puteal*, which stands near the north portal. It is apparently of the same æra as that part of the church.

From the cathedral we went to the church of St. Taurinus. The proud abbey of the apostle and first bishop of the diocese retains few or no traces of its former dignity. So long as monachism flourished, a contest existed between the chapter of the cathedral and the brethren of this monastery, each advocating the precedency of their respective establishment.—The monks of St. Taurinus contended, that their abbey was expressly mentioned by William of Jumieges[42] among the most ancient in Neustria, as well as among those which were destroyed by the Normans, and rebuilt by the zeal of good princes. They also alleged the dispute that prevailed under the Norman dukes for more than two hundred years, between this convent and that of Fécamp, respecting the right of nominating one of their own brethren to the head of their community, a right which was claimed by Fécamp; and they displayed the series of their prelates, continued in an uninterrupted line from the time of their founder. Whatever may have been the justice of these claims, the antiquity of the monastery is admitted by all parties.—Its monks, like those of the abbey of St. Ouen, had the privilege of receiving every new bishop of the see, on the first day of his arrival at Evreux; and his corpse was deposited in their church, where the funeral obsequies were performed. This privilege, originally intended only as a mark of distinction to the abbey, was on two occasions perverted to a purpose that might scarcely have been expected. Upon the death of Bishop John d'Aubergenville in 1256, the monks resented the reformation which he had endeavoured to introduce into their order, by refusing to admit his body within their precinct; and though fined for their obstinacy, they did not learn wisdom by experience, but forty-three years afterwards shewed their hostility decidedly towards the remains of Geoffrey of Bar, a still more determined reformer of monastic abuses. Extreme was the licentiousness which prevailed in those days among the monks of St. Taurinus, and unceasing were the endeavors of the bishop to correct them. The contest continued during his life, at the close of which they not only shut their doors against his corpse, but dragged it from the coffin and gave it a public flagellation. So gross an act of indecency would in all probability be classed among the many scandalous tales invented of ecclesiastics, but that the judicial proceedings which ensued leave no doubt of its truth; and it was even recorded in the burial register of the cathedral.

The church of St. Taurinus offers some valuable specimens of ancient architecture.—The southern transept still preserves a row of Norman arches, running along the lower part of its west side, as well as along its front; but those above them are pointed. To the south are six circular arches, divided into two compartments, in each of which the central arch has formerly served for a window. Both the lateral ones are filled with coeval stone-work, whose face is carved into lozenges, which were alternately coated with blue and red mortar or stucco: distinct traces of the coloring are still left in the cavities[43]. To the eastern side of this transept is attached, as at St. Georges, a small chapel, of semi-circular architecture, now greatly in ruins. The interior of the church is all comparatively modern, with the exception of some of the lower arches on the north side.—A strange and whimsical vessel for holy water attracted our attention. I cannot venture to guess at its date, but I do not think it is more recent than the fourteenth century.

The principal curiosity of the church, and indeed of the town, is the shrine, which contained, or perhaps, contains, a portion of the bones of the patron saint, whose body, after having continued for more than three hundred years a hidden treasure, was at last revealed in a miraculous manner to the prayers of Landulphus, one of his successors in the episcopacy.—The cathedral of Chartres, in early ages, set up a rival claim for the possession of this precious relic; but its existence here was formally verified at the end of the seventeenth century, by the opening of the châsse, in which a small quantity of bones was found tied up in a leather bag, with a certificate of their authenticity, signed by an early bishop.—The shrine is of silver-gilt, about one and a half foot in height and two feet in length: it is a fine specimen of ancient art. In shape it resembles the nave of a church, with the sides richly enchased with figures of saints and bishops. Our curious eyes would fain have pried within; but it was closed with the impression of the archbishop's signet.—A crypt, the original burial place of St. Taurinus, is still shewn in the church, and it continues to be the object of great veneration. It is immediately in front of the high altar, and is entered by two staircases, one at the head, the other at the foot of the coffin. The vault is very small, only admitting of the coffin and of a narrow passage by its side. The sarcophagus, which is extremely shallow, and neither wide nor long, is partly imbedded in the wall, so that the head and foot and one side alone are visible.—A portion of the monastic buildings of St. Taurinus now serves as a seminary for the catholic priesthood.

The west front of the church of St. Giles is not devoid of interest. Many other churches here have been desecrated; and this ancient building has been converted into a stable. The door-way is formed by a fine semi-circular arch, ornamented with the chevron-moulding, disposed in a triple row, and with a line of quatrefoils along the archivolt. Both these decorations are singular: I recollect no other instance of the quatrefoil being employed in an early Norman building, though immediately upon the adoption of the pointed style it became exceedingly common; nor can I point out another example of the chevron-moulding thus disposed. It produces a better effect than when arranged in detached bands. The capitals to the pillars of the arch are sculptured with winged dragons and other animals, in bold relief.

These are the only worthy objects of architectural inquiry now existing in the city. Many must have been destroyed by the ravages of war, and by the excesses of the revolution.—Evreux therefore does not abound with memorials of its antiquity. But its existence as a town, during the period of the domination of the Romans, rests upon authority that is scarcely questionable. It has been doubted whether the present city, or a village about three miles distant, known by the name of Old Evreux, is the Mediolanum

Aulercorum of Ptolemy. His description is given with sufficient accuracy to exclude the pretensions of any other town, though not with such a degree of precision as will enable us, after a lapse of sixteen centuries, to decide between the claims of the two sites. Cæsar, in his Commentaries, speaks in general terms of the Aulerci Eburovices, who are admitted to have been the ancient inhabitants of this district, and whose name, especially as modified to Ebroici and Ebroi, is clearly to be recognized in that of the county. The foundations of ancient buildings are still to be seen at Old Evreux; and various coins and medals of the upper empire, have at different times been dug up within its precincts. Hence it has been concluded, that the Mediolanum Aulercorum was situated there. The supporters of the contrary opinion admit that Old Evreux was a Roman station; but they say that, considering its size, it can have been no more than an encampment: they also maintain, that a castle was subsequently built upon the site of this encampment, by Richard, Count of Evreux, and that the destruction of this castle, during the Norman wars, gave rise to the ruins now visible, which in their turn were the cause of the name of the village[44].

It is certain that, in the reign of William the Conqueror, the town stood in its present situation: Ordericus Vitalis speaks in terms that admit of no hesitation, when he states that, in the year 1080, "fides Christi Evanticorum, id est Evroas, urbem, super Ittonum fluvium sitam possidebat et salubritèr illuminabat[45]."

In the times of Norman sovereignty, Evreux attained an unfortunate independence: Duke Richard Ist severed it from the duchy, and erected it into a distinct earldom in favor of Robert, his second son. From him the inheritance descended to Richard and William, his son and grandson; after whose death, it fell into the female line, and passed into the house of Montfort d'Amaury, by the marriage of Agnes, sister of Richard of Evreux.—Nominally independent, but really held only at the pleasure of the Dukes of Normandy, the rank of the earldom occasioned the misery of the inhabitants, who were continually involved in warfare, and plundered by conflicting parties. The annals of Evreux contain the relation of a series of events, full of interest and amusement to us who peruse them; but those, who lived at the time when these events were really acted, might exclaim, like the frogs in the fable, "that what is entertainment to us, was death to them."—At length, the treaty of Louviers, in 1195, altered the aspect of affairs. The King of France gained the right of placing a garrison in Evreux; and, five years afterwards, he obtained a formal cession of the earldom. Philip Augustus took possession of the city, to the great joy of the inhabitants, who, six years before, had seen their town pillaged, and their houses destroyed, by the orders of this monarch. The severity exercised upon that occasion had been excessive; but Philip's indignation had been roused by one of the basest acts of treachery recorded in history.—John, faithless at every period of his life, had entered into a treaty with the French monarch, during the captivity of his brother, Coeur-de-Lion, to deliver up Normandy; and Philip, conformably with this plan, was engaged in reducing the strong holds upon the frontiers, whilst his colleague resided at Evreux. The unexpected release of the English king disconcerted these intrigues; and John, alarmed at the course which he had been pursuing, thought only how to avert the anger of his offended sovereign. Under pretence, therefore, of shewing hospitality to the French, he invited the principal officers to a feast, where he caused them all to be murdered; and he afterwards put the rest of the garrison to the sword.—Brito records the transaction in the following lines, which I quote, not only as an historical document, illustrative of the moral character of one of the worst sovereigns that ever swayed the British sceptre, but as an honorable testimony to the memory of his unfortunate brother:—

"Attamen Ebroïcam studio majore reformans
Armis et rebus et bellatoribus urbem,
Pluribus instructam donavit amore Johanni,
Ut sibi servet eam: tamen arcem non dedit illi.
Ille dolo plenus, qui patrem, qui modo fratrem
Prodiderat, ne non et Regis proditor esset,
Excedens siculos animi impietate Tyrannos,
Francigenas omnes vocat ad convivia quotquot
Ebroïcis reperit, equites simul atque clientes,
Paucis exceptis quos sors servavit in arce.
Quos cum dispositis armis fecisset ut una
Discubuisse domo, tanquam prandere putantes,
Evocat e latebris armatos protinus Anglos,
Interimitque viros sub eadem clade trecentos,
Et palis capita ambustis affixit, et urbem
Circuit affixis, visu mirabile, tali
Regem portento quærens magis angere luctu:
Talibus obsequiis, tali mercede rependens
Millia marcharum, quas Rex donaverat illi.
Tam detestanda pollutus cæde Johannes
Ad fratrem properat; sed Rex tam flagitiosus
Non placuit fratri: quis enim, nisi dæmone plenus,

Omninoque Deo vacuus, virtute redemptus
A vitiis nulla, tam dira fraude placere
Appetat, aut tanto venetur crimine pacem?
Sed quia frater erat, licet illius oderit actus
Omnibus odibiles, fraternæ foedera pacis
Non negat indigno, nec eum privavit amore,
Ipsum qui nuper Regno privare volebat."

The vicissitudes to which the county of Evreux was doomed to be subject, did not wholly cease upon its annexation to the crown of France. It passed, in the fourteenth century, into the hands of the Kings of Navarre, so as to form a portion of their foreign territory; and early in the fifteenth, it fell by right of conquest under English sovereignty.—Philip the Bold conferred it, in 1276, upon Louis, his youngest son; and from him descended the line of Counts of Evreux, who, originating in the royal family of France, became Kings of Navarre. The kingdom was brought into the family by the marriage of Philip Count of Evreux with Jane daughter of Louis Hutin, King of France and Navarre, to whom she succeeded as heir general. Charles IIIrd, of Navarre, ceded Evreux by treaty to his namesake, Charles VIth of France, in 1404; and he shortly after bestowed it upon John Stuart, Lord of Aubigni, and Constable of Scotland.—Under Henry Vth, our countrymen took the city in 1417, but we were not long allowed to hold undisturbed possession of it; for, in 1424, it was recaptured by the French. Their success, however, was only ephemeral: the battle of Verneuil replaced Evreux in the power of the English before the expiration of the same year; and we kept it till 1441, when the garrison was surprised, and the town lost, though not without a vigorous resistance.—Towards the close of the following century, the earldom was raised into a Duché pairie, by Charles IXth, who, having taken the lordship of Gisors from his brother, the Duc d'Alençon, better known by his subsequent title of Duc d'Anjou, recompenced him by a grant of Evreux. Upon the death of this prince without issue, in 1584, Evreux reverted to the crown, and the title lay dormant till 1652, when Louis XIVth exchanged the earldom with the Duc de Bouillon, in return for the principality of Sedan. In his family it remained till the revolution, which, amalgamating the whole of France into one common mass of equal rights and laws, put an end to all local privileges and other feudal tenures.

Evreux, at present, is a town containing about eight thousand inhabitants, a great proportion of whom are persons of independent property, or rentiers, as the French call them. Hence it has an air of elegance, seldom to be found in a commercial, and never in a manufacturing town; and to us this appearance was the more striking, as being the first instance of the kind we had seen in Normandy. The streets are broad and beautifully neat. The city stands in the midst of gardens and orchards, in a fertile valley, watered by the Iton, and inclosed towards the north and south by ranges of hills. The river divides into two branches before it reaches the town, both which flow on the outside of the walls. But, besides these, a portion of its waters has been conducted through the centre of the city, by means of a canal dug by the order of Jane of Navarre. This Iton, like the Mole, in Kent, suddenly loses itself in the ground, near the little town of Damville, about twenty miles south of Evreux, and holds its subterranean course for nearly two miles. A similar phenomenon is observable with a neighboring stream, the Risle, between Ferrière and Grammont[46]: in both cases it is attributed, I know not with what justice, to an abrupt change in the stratification of the soil.

Footnotes:

[36] This curious transaction, which took place in the year 1119, is related with considerable näiveté by Ordericus Vitalis, p. 852, as follows:—"Henricus Rex rebellibus ultrà parcere nolens, pagum Ebroicensem adiit, et Ebroas cum valida manu impugnare coepit. Sed oppidanis, qui intrinsecus erant, cum civibus viriliter repugnantibus, introire nequivit. Erant cum illo Ricardus filius ejus, et Stephanus Comes nepos ejus, Radulfus de Guader, et maxima vis Normannorum. Quibus ante Regem convocatis in unnm, Rex dixit ad Audinum Episcopum. "Videsne, domine Præsul, quòd repellimur ab hostibus, nec eos nisi per ignem subjugare poterimus? Verùm, si ignis immittitur, Ecclesiæ comburentur, et insontibus ingens damnum inferetur. Nunc ergo, Pastor Ecclesiæ, diligentèr considera, et quod utilius prospexeris providè nobis insinua. Si victoria nobis per incendium divinitùs conceditur, opitulante Deo, Ecclesiæ detrimenta restaurabuntur: quia de thesauris nostris commodos sumptus gratantèr largiemur. Unde domus Dei, ut reor, in melius reædificabuntur." Hæsitat in tanto discrimine Præsul auxius, ignorat quid jubeat divinæ dispositioni competentius: nescit quid debeat magis velle vel eligere salubrius. Tandem prudentum consultu præcepit ignem immitti, et civitatem concremari, ut ab anathematizatis proditoribus liberaretur, et legitimis habitatoribus restitueretur. Radulfus igitur de Guader a

parte Aquilonali primus ignem injecit, et effrenis flamma per urbem statim volavit, et omnia (tempos enim autumni siccum erat) corripuit. Tunc combusta est basilica sancti Salvatoris, quam Sanctimoniales incolebant, et celebris aula gloriosæ virginis et matris Mariæ, cui Præsul et Clerus serviebant, ubi Pontificalem Curiam parochiani frequentabant. Rex, et cuncti Optimales sui Episcopo pro Ecclesiarum combustione vadimonium supplicitèr dederunt, et uberes impensas de opibus suis ad restaurationem earum palam spoponderunt."

[37] Duchesne, Scriptores Normanni, p. 309.

[38] Gallia Christiana, XI. p. 606.

[39] From the manner in, which Ducarel speaks of these statues, (Anglo-Norman Antiquities, p. 85.) he leaves it to be understood, that they were in existence in his time; but it is far from certain that this was the case; for the whole of his account of them is no more than a translation from the following passage in Le Brasseur's Histoire du Comté d'Evreux, p. 11.—"Le Diocèse d'Evreux a été si favorisé des grâces de Dieu, qu'on ne voit presqu'aucun temps où l'Hérésie y ait pénétré, même lorsque les Protestans inondoient et corrompoient toute la France, et particulierement la Normandie. On ne peut pas cependant desavoüer qu'il y a eu de temps en temps, quelques personnes qui se sont livrées à l'erreur; et l'on peut remarquer quatre Statuës attachées à deux piliers au dehors du chancel de l'Eglise Cathédrale du côté du Cimetiere, dont trois représentent trois Chanoines, la tête couverte de leurs Aumuces selon la coûtume de ce temps-là, et une quatrième qui représente un Chanoine à un pilier plus éloigné, la tête nuë, tenant sa main sur le coeur comme un signe de son repentir; parce que la tradition dit, qu'aïant été atteint et convaincu du crime d'hérésie, le Chapitre l'avoit interdit des fonctions de son Bénéfice; mais qu'aïant ensuite abjuré son erreur, le même Chapitre le rétablit dans tous ses droits, honneurs, et privileges: cependant il fut ordonné qu'en mémoire de l'égarement et de la pénitence de ce Chanoine, ces Statuës demeureroient attachées aux piliers de leur Eglise, lorsqu'elle fût rébâtie des deniers de Henry I. Roy d'Angleterre, par les soins d'Audoenus Evêque d'Evreux."

[40] This was not the first, nor the only, contest, which was fought by Taurinus with Satan. Their struggles began

at the moment of the saint's coming to Evreux, and did not even terminate when his life was ended. But the devil was, by the power of his adversary, brought to such a helpless state, that, though he continued to haunt the city, where the people knew him by the name of Gobelinus, he was unable to injure any one.—All this is seriously related by Ordericus Vitalis, (p. 555.) from whom I extract the following passage, in illustration of what Evreux was supposed to owe to its first bishop.—"Grassante secundâ persecutione, quæ sub Domitiano in Christianos furuit, Dionysius Parisiensis Episcopus Taurinum filiolum suum jam quadragenarium, Præsulem ordinavit; et (vaticinatis pluribus quæ passurus erat) Ebroicensibus in nomine Domini direxit. Viro Dei ad portas civitatis appropinquanti, dæmon in tribus figmentis se opposuit: scilicet in specie ursi, et leonis, et bubali terrere athletam Christi voluit. Sed ille fortiter, ut inexpugnabilis murus, in fide perstitit, et coeptum iter peregit, hospitiumque in domo Lucii suscepit. Tertia die, dum Taurinus ibidem populo prædicaret, et dulcedo fidei novis auditoribus multùm placeret, dolens diabolus Eufrasiam Lucii filiam vexare coepit, et in ignem jecit. Quæ statim mortua est; sed paulò pòst, orante Taurino ac jubente ut resurgeret, in nomine Domini resuscitata est. Nullum in ea adustionis signum apparuit. Omnes igitur hoc miraculum videntes subitò territi sunt, et obstupescentes in Dominum Jesum Christum crediderunt. In illa die c. homines baptizati sunt. Octo cæci illuminati, et quatuor multi sanati, aliique plures ex diversis infirmitatibus in nomine Domini sunt curati."

[41] Masson de St. Amand, Essais Historiques sur Evreux, I. p. 77.

[42] Duchesne, Scriptores Normanni, p. 279.

[43] For this observation, as well as for several others touching Evreux and Pont-Audemer, I have to express my acknowledgments to Mr. Cotman's memoranda.

[44] Le Brasseur, Histoire du Comté d'Evreux, p. 4.

[45] Duchesne, Scriptores Normanni, p. 555.

[46] Goube, Histoire du Duché de Normandie, III. p. 223.

LETTER XIX
VICINITY OF EVREUX—CHÂTEAU DE NAVARRE—COCHEREL—PONT-AUDEMER —MONTFORT-SUR-RISLE—HARFLEUR— BOURG-ACHARD—FRENCH WEDDING

(Bourg-Achard, July, 1818.)

Evreux is seldom visited by the English; and none of our numerous absentees have thought fit to settle here, though the other parts of Normandy are filled with families who are suffering under the sentence of self-banishment. It is rather surprising, that this town has not obtained its share of English settlers: the air is good, provisions are cheap, and society is agreeable. Those, too, if such there be, who are attracted by historical reminiscences, will find themselves on historical ground.

The premier viscount of the British parliament derives his name from Evreux; though, owing to a slight alteration in spelling and to our peculiar pronunciation, it has now become so completely anglicised, that few persons, without reflection, would recognize a descendant of the Comtes d'Evreux, in Henry Devereux, Viscount of Hereford. The Norman origin of this family is admitted by the genealogists and heralds, both of France and of England; and the fate of the Earl of Essex is invariably introduced in the works of those authors, who have written upon Evreux or its honors.

It would have been unpardonable to have quitted Evreux, without rambling to the Château de Navarre, which is not more than a mile and half distant from the town.—This Château, whose name recals an interesting period in the history of the earldom, was originally a royal residence. It was erected in the middle of the fourteenth century by Jane of France, who, with a very pardonable vanity, directed her new palace to be called Navarre, that her Norman subjects might never forget that she was herself a queen, and that she had brought a kingdom as a marriage portion to her husband. Her son, Charles the Bad, a prince whose turbulent and evil disposition caused so much misfortune to France, was born here. Happy too had it been for him, had he here closed his eyes before he entered upon the wider theatre of the world! During his early days passed at Navarre, he is said to have shewn an ingenuousness of disposition and some traits of generosity, which gave rise to hopes that were miserably falsified by his future life.—The present edifice, however, a modern French Château, retains nothing more than the name of the structure which was built by the queen, and which was levelled with the ground, in the year 1686, by the Duc de Bouillon, the lord of the country, who erected the present mansion. His descendants resided here till the revolution, at which time they emigrated, and the estate became national property. It remained for a considerable period unoccupied, and was at last granted to Joséphine, by her imperial husband. At present, the domain belongs to her son, Prince Eugene, by whom the house has lately been stripped of its furniture. Many of the fine trees in the park have also been cut down, and the whole appears neglected and desolate. His mother did not like Navarre: he himself never saw it: the queen of Holland alone used occasionally to reside here.—The principal beauty of the place lies in its woods; and these we saw to the greatest advantage. It was impossible for earth or sky to look more lovely.—The house is of stone, with large windows; and an ill-shaped dome rises in the centre. The height of the building is somewhat greater than its width, which makes it appear top-heavy; and every thing about it is formal; but the noble avenue, the terrace-steps, great lanthorns, iron gates, and sheets of water on either side of the approach, are upon an extensive scale, and in a fine baronial style.—Yet, still they are inferior to the accompaniments of the same nature which are found about many noblemen's residences in England.—The hall, which is spacious, has a striking effect, being open to the dome. Its sides are painted with military trophies, and with the warlike instruments of the four quarters of the globe. We saw nothing else in the house worthy of notice. It is merely a collection of apartments of moderate size; and, empty and dirty as they were, they appeared to great disadvantage. In the midst of the solitude of desolation, some ordinary portraits of the Bouillon family still remain upon the walls, as if in mockery of departed greatness.

We were unable to direct our course to Cocherel, a village about sixteen miles distant, on the road to Vernon, celebrated as the spot where a battle was fought, in the fourteenth century, between the troops of Navarre, and those of France, commanded by Du Guesclin.—I notice this place, because it is possible that, if excavations were made there, those antiquaries who delight in relics of the remotest age of European history, might win many prizes. A tomb of great curiosity was discovered in the year 1685; and celts, and stone hatchets, and other implements, belonging, as it is presumed, to the original inhabitants of the country, have been found beneath the soil. Many of these are described and figured by the Abbé de Cocherel, in a paper full of curious erudition, subjoined to Le Brasseur's History of

Evreux. The hatchets resembled those frequently dug up in England; but they were more perfect, inasmuch as some of them were fastened in deers' horns, and had handles attached to them; thus clearly indicating the manner in which they were used.—The place of burial differed, I believe, in its internal arrangement from any sepulchral monument, whether Cromlech, Carnedd, or Barrow, that has been opened in our own country. Three sides of it were rudely faced with large stones: within were contained about twenty skeletons, lying in a row, close to each other, north and south, their arms pressed to their sides. The head of each individual rested on a stone, fashioned with care, but to no certain pattern. Some were fusiform, others wedge-shaped, and others irregularly oblong. In general, the stones did not appear to be the production of the country. One was oriental jade, another German agate. In the tomb were also a few cinerary urns; whence it appears that the people, by whom it was constructed, were of a nation that was at once in the habit of burning, and of interring, their dead. From these facts, the Abbé finds room for much ingenious conjecture; and, after discussing the relative probabilities of the sepulchre having been a burying-place of the Gauls, the Jews, the Druids, the Normans, or the Huns, he decides, though with some hesitation, in favor of the last of these opinions.

From Evreux we went by Brionne to Pont-Audemer: at first the road is directed through an open country, without beauty or interest; but the prospect improved upon us when we joined the rapid sparkling *Risle*, which waters a valley of great richness, bounded on either side by wooded hills.—Of Brionne itself I shall soon have a better opportunity of speaking; as we purpose stopping there on our way to Caen.

A few miles before Brionne, we passed Harcourt, the ancient barony of the noble family still flourishing in England, and existing in France. It is a small country town, remarkable only for some remains of a castle[47], built by Robert de Harcourt, fifth in descent from Bernard the Dane, chief counsellor, and second in command to Rollo. The blood of the Dane is in the present earl of Harcourt: he traces his lineage in a direct line from Robert, the builder of the castle, who accompanied the Conqueror into England, and fell in battle by his side.

Pont-Audemer is a small, neat, country town, situated upon the Risle, which here, within ten miles of its junction with the Seine, is enlarged into a river of considerable magnitude. But its channel, in the immediate vicinity of the town, divides into several small streams; and thus it loses much of its dignity, though the change is highly advantageous to picturesque beauty, and to the conveniences of trade. Mills stand on some of these streams, but most of them are applied to the purposes of tanning; for leather is the staple manufacture of the place, and the hides prepared at Pont-Audemer are thought to be the best in France.

From Brionne the valley of the Risle preserves a width of about a mile, or a mile and half: at Pont-Audemer it becomes somewhat narrower, and the town stretches immediately across it, instead of being built along the banks of the river.—The inhabitants are thus enabled to avail themselves of the different streams which intersect it.

Tradition refers the origin, as well as the name of Pont-Audemer, to a chief, called Aldemar or Odomar, who ruled over a portion of Gaul in the fifth century, and who built a bridge here.—These legendary heroes abound in topography, but it is scarcely worth while to discuss their existence. In Norman times Pont-Audemer was a military station. The nobility of the province, always turbulent, but never more so than during the reign of Henry Ist, had availed themselves of the opportunity afforded by the absence of the monarch, and by his domestic misfortunes, to take up arms in the cause of the son of Robert. Henry landed at the mouth of the Seine, and it was at Pont-Audemer that the first conflict took place between him and his rebellious subjects. The latter were defeated, and the fortress immediately surrendered; but, in the early part of the fourteenth century, it appears to have been of greater strength: it had been ceded by King John of France to the Count of Evreux, and it resisted all the efforts of its former lord during a siege of six weeks, at the end of which time his generals were obliged to retire, with the loss of their military engines and artillery. This siege is memorable in history, as the first in which it is known that cannon were employed in France.—Pont-Audemer, still in possession of the kings of Navarre, withstood a second siege, towards the conclusion of the same century, but with less good fortune than before. It was taken by the constable Du Guesclin, and, according to Froissart[48], "the castle was razed to the ground, though it had cost large sums to erect; and the walls and towers of the town were destroyed."

St. Ouen, the principal church in the place, is a poor edifice. It bears, however, some tokens of remote age: such are the circular arches in the choir, and a curious capital, on which are represented two figures in combat, of rude sculpture.—A second church, that of Notre Dame des Prés, now turned into a tan-house, exhibits an architectural feature which is altogether novel. Over the great entrance, it has a string-course, apparently intended to represent a corbel-table, though it does not support any superior member; and the intermediate spaces between the corbels, instead of being left blank, as usual, are filled with sculptured stones, which project considerably, though less than the corbels with which they alternate. There is something of the same kind, but by no means equally remarkable, over the arcades above the west door-way of Castle-Acre Priory[49]. Neither Mr. Cotman's memory, nor my own, will furnish another example.—The church of Notre Dame des Prés is of the period when the pointed style was beginning to be employed. The exterior is considerably injured: to the interior we could not obtain admission.

The suburbs of Pont-Audemer furnish another church dedicated to St. Germain, which would have been an excellent subject for both pen and pencil, had it undergone less alteration. The short, thick, square, central tower has, on each side, a row of four windows, of nearly the earliest pointed style; many of the windows of the body of the church have semi-circular heads; the corbels which extend in a line round the nave and transepts are strangely grotesque; and, on the north side of the eastern extremity, is a semi-circular chapel, as at St. Georges.—The inside is dark and gloomy, the floor unpaved, and every thing in and about it in a state of utter neglect, except some dozen saints, all in the gayest attire, and covered with artificial flowers. The capitals of the columns are in the true Norman style. Those at St. Georges are scarcely more fantastic, or more monstrous.—Between two of the arches of the choir, on the south side of this church, is the effigy of a man in his robes, coifed with a close cap, lying on an altar-tomb. The figure is much mutilated; but the style of the canopy-work over the head indicates that it is not of great antiquity. The feet of the statue rest upon a dog, who is busily occupied in gnawing a marrow-bone.—Dogs at the base of monumental effigies are common, and they have been considered as symbols of fidelity and honor; but surely the same is not intended to be typified by a dog thus employed; and it is not likely that his being so is a mere caprice of the sculptor's.—There is no inscription upon the monument; nor could we learn whom it is intended to commemorate.

At but a short distance from Pont-Audemer, higher up the Risle, lies the yet smaller town of Montfort, near which are still to be traced, the ruins of a castle,[50] memorable for the thirty days' siege, which it supported from the army of Henry Ist, in 1122; and dismantled by Charles Vth, at the same time that he razed the fortifications of Pont-Audemer. The Baron of Montfort yet ranks in our peerage; though I am not aware that the nobleman, who at present bears the title, boasts a descent from any part of the family of Hugh with a beard, the owner of Montfort at the time of the conquest, and one of the Conqueror's attendants at the battle of Hastings.

From Pont-Audemer we proceeded to Honfleur: it was market-day at the place which we had quitted, and the throng of persons who passed us on the road, gave great life and variety to the scene. There was scarcely an individual from whom we did not receive a friendly smile or nod, accompanied by a *bon jour*; for the practice obtains commonly in France, among the peasants, of saluting those whom they consider their superiors. Almost all that were going to market, whether male or female, were mounted on horses or asses; and their fruit, vegetables, butchers' meat, live fowls, and live sheep, were indiscriminately carried in the same way.

About a league before we arrived at Honfleur, a distant view of the eastern banks of the river opened upon us from the summit of a hill, and we felt, or fancied that we felt, "the air freshened from the wave." As we descended, the ample Seine, here not less than nine miles in width, suddenly displayed itself, and we had not gone far before we came in sight of Honfleur. The mist occasioned by the intense heat, prevented us from seeing distinctly the opposite towns of Havre and Harfleur: we could only just discern the spire of the latter, and the long projecting line of the piers and fortifications of Havre. The great river rolls majestically into the British Channel between these two points, and forms the bay of Honfleur. About four miles higher up the stream where it narrows, the promontories of Quilleboeuf and of Tancarville close the prospect.—Honfleur itself is finely situated: valleys, full of meadows of the liveliest green, open to the Seine in the immediate vicinity of the town; and the hills with which it is backed are beautifully clothed with foliage to the very edge of the water. The trees, far from being stunted and leafless, as on the eastern coast of England, appear as if they were indebted to their situation for a verdure of unusual luxuriancy. A similar line of hills borders the Seine on either side, as far as the eye can reach.

It was unfortunate for us, that we entered the town at low water, when the empty harbor and slimy river could scarcely fail to prepossess us unfavorably. The quays are faced with stone, and the two basins are fine works, and well adapted for commerce. This part of Honfleur reminded us of Dieppe; but the houses, though equally varied in form and materials, are not equally handsome.—Still less so are the churches; and a picturesque castle is wholly wanting.—In the principal object of my journey to Honfleur, my expectations were completely frustrated. I had been told at Rouen, that I should here find a very ancient wooden church, and our imagination had pictured to us one equally remarkable as that of Greensted, in Essex, and probably constructed in the same manner, of massy trunks of trees. With the usual anticipation of an antiquary, I imagined that I should discover a parallel to that most singular building; which, as every body knows, is one of the greatest architectural curiosities in England. But, alas! I was sadly disappointed. The wooden church of Honfleur, so old in the report of my informant, is merely a thing of yesterday, certainly not above two hundred and fifty years of age; and, though it is undeniably of wood, within and without, the walls are made, as in most of the houses in the town, of a timber frame filled with clay. There is another church in Honfleur, but it was equally without interest. Thus baffled, we walked to the heights above the town: at the top of the cliff was a crowd of people, some of them engaged in devotion near a large wooden crucifix, others enjoying themselves at different games, or sitting upon the neat stone benches, which are scattered plentifully about the walks in this charming situation. The neighboring little chapel of Notre Dame de Grace is regarded as a building of great sanctity, and is especially resorted to by sailors, a class of people who are superstitious, all the world over. It abounds with their votive tablets. From the roof and walls

> "Pendono intorno in lungo ordine i voti,
> Che vi portaro i creduli divoti."

Among the pictures, we counted nineteen, commemorative of escape from shipwreck, all of them painted after precisely the same pattern: a stormy sea, a vessel in distress, and the Virgin holding the infant Savior in her arms, appearing through a black cloud in the corner,—In the Catholic ritual, the holy Virgin, is termed Maris Stella, and she is κατ' εξοχην [English. Not in Original: pre-eminently, especially, above all] the protectress of Normandy.

Honfleur is still a fortified town; but it does not appear a place of much strength, nor is it important in any point of view. Its trade is inconsiderable, and its population does not amount to nine thousand inhabitants. But in the year 1450, while in the hands of our countrymen, it sustained a siege of a month's duration from the king of France; and, in the following century, it had the distinction, attended with but little honor, of being the last place in the kingdom that held out for the league.

From Honfleur we would fain have returned by Sanson-sur-Risle and Foullebec, at both which villages M. Le Prevost had led us to expect curious churches; but our postillion assured us that the roads were wholly impassable. We were therefore compelled to allow Mr. Cotman to visit them alone, while we retraced a portion of our steps through the valley of the Risle, and then took an eastern direction to Bourg-Achard in our way to Rouen.

Bourg-Achard was the seat of an abbey, built by the monks of Falaise, in 1143: it was originally dedicated to St. Lô; but St. Eustatius, the favorite saint of this part of the country, afterwards became its patron. Before the revolution, his skull was preserved in the sacristy of the convent, enchased in a bust of silver gilt[51]; and even now, when the relic has been consigned to its kindred dust, and the shrine to the furnace, and the abbey has been levelled with the ground, there remains in the parochial church a fragment of sculpture, which evidently represented the miracle that led to Eustatius' conversion.—The knight, indeed, is gone, and the cross has disappeared from between the horns of the stag; but the horse and the deer, are left, and their position indicates the legend.—The church of Bourg-Achard has been materially injured. The whole of the building, from the transept westward, has been taken down; but it deserves a visit, if only as retaining a bénitier of ancient form and workmanship, and a leaden font. Of the latter, I send you a drawing. Leaden fonts are of very rare occurrence in England[52], and I never saw or heard of another such in France: indeed, a baptismal font of any kind is seldom to be seen in a French church, and the vessels used for containing the holy water, are in most cases nothing more than small basins in the form of escalop shells, affixed to the wall, or to some pillar near the entrance.—It is possible that the fonts were removed and sold during the revolution, as they were in our own country, by the ordinance of the houses of parliament, after the deposition of Charles Ist; but this is a mere conjecture on my own part. It is also possible that they may be kept in the sacristy, where I have certainly seen them in some cases. In earlier times, they not only existed in every church, but were looked upon with superstitious reverence. They are frequently mentioned in the decrees of ecclesiastical councils; some of which provide for keeping them clean and locked; others for consigning the keys of them to proper officers; others direct that they should never be without water; and others that nothing profane should be laid upon them[53].

As we were at breakfast this morning, a procession, attended by a great throng, passed our windows, and we were invited by our landlady to go to the church and see the wedding of two of the principal persons of the parish, We accepted the proposal; and, though the same ceremony has been witnessed by thousands of Englishmen, yet I doubt whether it has been described by any one.—The bride was a girl of very interesting appearance, dressed wholly in white: even her shoes were white, and a bouquet of white roses, jessamine, and orange-flowers, was placed in her bosom.—The mayor of the town conducted her to the altar. Previously to the commencement of the service, the priest stated aloud that the forms required by law, for what is termed the civil marriage, had been completed. It was highly necessary that he should do so; for, according to the present code, a minister of any persuasion, who proceeds to the religious ceremonies of marriage before the parties have been married by the magistrate, is subject to very heavy penalties, to imprisonment, and to transportation. Indeed, going to church at all for the purpose of marriage, is quite a work of supererogation, and may be omitted or not, just as the parties please; the law requiring no other proof of a marriage, beyond the certificate recorded in the municipal registry. After this most important preliminary, the priest exhorted every one present, under pain of excommunication, to declare if they knew of any impediment: this, however, was merely done for the purpose of keeping up the dignity of the church, for the knot was already tied as fast as it ever could be. He then read a discourse upon the sanctity of the marriage compact, and the excellence of the wedded state among the Catholics, compared to what prevailed formerly among the Jews and Heathens, who degraded it by frequent divorces and licentiousness. The parties now declared their mutual consent, and his reverence enjoined each to be to the other "comme un époux fidèle et de lui tenir fidélité en toutes choses."—The ring was presented to the minister by one of the acolytes, upon a gold plate; and, before he directed the bridegroom to place it upon the finger of the lady, he desired him to observe that it was a symbol of marriage.—During the whole of the service two other acolytes were stationed in front of the bride and bridegroom, each holding in his hands a lighted taper; and near the conclusion, while they knelt before the altar, a pall of flowered brocade was stretched behind them, as emblematic of their union. Holy water was not forgotten; for, in almost every rite of the Catholic church, the mystic sanctification by water and by fire continually occurs.—The ceremony ended by the priest's receiving the sacrament himself, but without administering it to any other individual present. Having taken it, he kissed the paten which had contained the holy elements, and all the party did the same: each, too, in succession, put a piece of money into a cup, to which we also were invited to contribute, for the love of the Holy Virgin.—They entered by the south door, but the great western portal was thrown open as they left the church; and by that they departed.

Footnotes:

[47] Masson de St. Amand, Essais Historiques sur Evreux, I. p. 39.

[48] Johnes' Translation, 8vo, IV. p. 292.

[49] See Britten's Architectural Antiquities, III. t. 2.

[50] Goube, Histoire de Normandie, III. 249.

[51] Histoire de la Haute Normandie, II. p. 319.

[52] Mr. Gough, (See Archæologia, X. p. 187.) whose attention had been much directed to this subject, seems to have known only four fonts made of lead, in the kingdom;—at Brookland in Kent, Dorchester in Oxfordshire, Wareham in Dorsetshire, and Walmsford in Northamptonshire; but there are in all probability many more. We have at least four in Norfolk. He says, "they are supposed to be of high antiquity; and that at Brookland may have relation to the time of Birinus himself. To what circumstance the others are to be referred, or from what other church brought, does not appear."—The leaden fonts which I have seen, have all been raised upon a basis of brick or stone, like this at Bourg-Achard, and are all of nearly the same pattern.

[53] See Concilia Normannica, II. pp. 56, 117, 403, 491, 508, &c.

LETTER XX
MOULINEAUX—CASTLE OF ROBERT THE DEVIL—BOURG-THEROUDE—ABBEY OF BEC—BRIONNE

(*Brionne, July*, 1818.)

Having accomplished the objects which we had proposed to ourselves in Rouen and its vicinity, we set out this morning upon our excursion to the western parts of the province. Our first stage, to Moulineaux, was by the same road by which we returned a few days ago from Bourg-Achard. It is a delightful ride, through the valley of the Seine, here of great width, stretching to our left in an uninterrupted course of flat open country, but, on our right hand, bordered at no great distance by the ridge of steep chalky cliffs which line the bank of the river. The road appears to have been a work of considerable labor: it is every where raised, and in some places as high as fifteen feet above the level of the fields on either side.—Agriculture in this district is conducted, as about Paris, upon the plan called by the French la petite culture: the fields are all divided into narrow strips; so that a piece of not more than two or three acres, frequently produces eight or ten different crops, some of grain, others of culinary vegetables, at the same time that many of these portions are planted with apple and cherry trees. The land is all open and uninclosed: not a fence is to be seen; nor do there even appear to be any balks or head-marks. Strangers therefore who come, like us, from a country entirely inclosed, cannot refrain from frequent expressions of surprise how it is that every person here is enabled to tell the limits of his own property.

Moulineaux is a poor village, a mere assemblage of cottages, with mud walls and thatched roofs. But the church is interesting, though desecrated and verging to ruin. Even now the outside alone is entire. The interior is gutted and in a state of absolute neglect.—The building is of the earliest pointed style: its lancet-windows are of the plainest kind, being destitute of side pillars: in some of the windows are still remains of handsome painted glass.—Either the antiquaries in France are more honest than in England, or they want taste, or objects of this kind do not find a ready market. We know too well how many an English church, albeit well guarded by the churchwardens and the parson, has seen its windows despoiled of every shield, and saint, and motto; and we also know full well, by whom, and for whom, such ravages are committed. In France, on the contrary, where painted glass still fills the windows of sacred buildings, now employed for the meanest purposes, or wholly deserted, no one will even take the trouble of carrying it away; and the storied panes are left, as derelicts utterly without value.—The east end of the church at Moulineaux is semi-circular; the roof is of stone, handsomely groined, and the groinings spring from fanciful corbels. On either side of the nave, near the choir, is a recess in the wall, carved with tabernacle-work, and serving for a piscina. Recesses of this kind, though of frequent occurrence in English churches, do not often appear in France. Still less common are those elaborate screens of carved timber, often richly gilt or gorgeously painted, which separate the nave from the chancel in the churches of many of our smaller villages at home. The only one I ever recollect to have seen in France was at Moulineaux.—I also observed a mutilated pillar, which originally supported the altar, ornamented with escalop shells and fleurs-de-lys in bold relief. It reminded me of one figured in the Antiquarian Repertory, from Harold's chapel, in Battle Abbey[54].

Immediately after leaving Moulineaux, the road winds along the base of a steep chalk hill, whose brow is crowned by the remains of the famous castle of Robert the Devil, the father of Richard Fearnought. Robert the Devil is a mighty hero of romance; but there is some difficulty in discovering his historical prototype. Could we point out his gestes in the chronicle, they would hardly outvalue his adventures, as they are recorded in the nursery tale. Robert haunts this castle, which appears to have been of great extent, though its ruins are very indistinct. The walls on the southern side are rents, and covered with brush-wood; and no architectural feature is discernible. Wide and deep fosses encircle the site, which is undermined by spacious crypts and subterraneous caverns.—The fortress is evidently of remote, but uncertain, antiquity: it was dismantled by King John when he abandoned the duchy. The historians of Normandy say that it was re-fortified during the civil wars; and the fact is not destitute of probability, as its position is bold and commanding.

Bourg-Theroude, our next stage, is one of those places which are indebted to their names alone for the little importance they possess. At present, it is a small assemblage of mean houses, most of them inns; but its Latin appellation, Burgus Thuroldi, commemorates no less a personage than one of the preceptors of William the Conqueror, and his grand constable at the time when he effected the conquest of England.—The name of Turold occurs upon the Bayeux tapestry, designating one of the ambassadors dispatched by the Norman Duke to Guy, Earl of Ponthieu; and it is supposed that the Turold there represented was the grand constable[55].—The church of Bourg-Theroude, which was collegiate before the revolution, is at present uninteresting in every point of view.

About half way from this place to Brionne, we came in sight of the remains of the celebrated abbey of Bec, situated a mile and half or two miles distant to our right, at the extremity of a beautiful valley. We had been repeatedly assured that scarcely one stone of this formerly magnificent building was left upon another; but it would have shewn an unpardonable want of curiosity to have passed so near without visiting it: even to stand upon the spot which such a monastery originally covered is a privilege not lightly to be foregone:—

"The pilgrim who journeys all day,
To visit some far distant shrine;
If he bear but a relic away,
Is happy, nor heard to repine."—

And *happiness* of this kind would on such an occasion infallibly fall to your lot and to mine. A love for botany or for antiquities would equally furnish *relics* on a similar *pilgrimage*.

As usual, the accounts which we had received proved incorrect. The greater part of the conventual edifice still exists, but it has no kind of architectural value. Some detached portions, whose original use it would be difficult now to conjecture, appear, from their wide pointed windows, to be of the fifteenth century. The other buildings were probably erected within the last fifty years.—The part inhabited by the monks is at this time principally employed as a cotton-mill; and, were it in England, nobody would suspect that it ever had any other destination. Of the church, the tower[56] only is in existence. I find no account of its date; though authors have been unusually profuse in their details of all particulars relating to this monastery. I am inclined to refer it to the beginning of the seventeenth century, in which case it was built shortly after the destruction of the nave. Its character is simple, solid elegance. Its ornaments are few, but they are selected and disposed with judgment. Each corner is flanked by two buttresses, which unite at top, and there terminate in a crocketed pinnacle. The buttresses are also ornamented with tabernacles of saints at different heights; and one of the tabernacles upon each buttress, about mid-way up the tower, still retains a statue as large as life, of apparently good workmanship. They were fortunately too high for the democrats to destroy with ease. The height of the tower is one hundred and fifty feet, as I found by the staircase of two hundred steps, which remains uninjured, in a circular turret attached to the south side. The termination of this turret is the most singular part of the structure: it is surmounted by a cap, considerably higher than the pinnacles, and composed, like a bee-hive, of a number of circles, each smaller than the one below it. A few ruined arches of the east end of the church, and of one of the side chapels are also existing. The rest is levelled with the ground, and has probably been in a great measure destroyed lately; for piles of wrought stones are heaped up on all sides.

If historical recollections or architectural beauty could have proved a protection in the days of revolution, the church of Bec had undoubtedly stood. Ducarel, who saw it in its perfection, says it was one of the finest gothic structures in France; and his account of it, though only an abridgement of that given by Du Plessis, in his History of Upper Normandy, is curious and valuable.—Mr. Gough states the annual income of the abbey at the period of the revolution, to have exceeded twenty thousand crowns. Its patronage was most extensive: the monks presented to one hundred and sixty advowsons, two of them in the metropolis; and thirty other ecclesiastical benefices, as well priories as chapels, were in their gift[57].—Its possessions, as we may collect from the various charters and donations, might have led us to expect a larger revenue. The estates belonging to the monastery in England, prior to the reformation, were both numerous and valuable.

Sammarthanus, author of the Gallia Christiana, says, in speaking of Bec, that, whether considered as to religion or literature, there was not, in the eleventh century, a more celebrated convent throughout the whole of Neustria. The founder of the abbey was Hellouin, sometimes called Herluin, a nobleman, descended by the mother's side from the Counts of Flanders, but he himself was a native of the territory of Brionne, and educated in the castle of Gislebert, earl of that district. Hellouin determined, at an early age, to withdraw himself from the court and from the world: it seems he was displeased or affronted by the conduct of the earl; and we may collect from the chroniclers, that it was not a very easy task in those times for an individual of rank, intent upon monastic seclusion, to carry his purpose into effect, and that still greater difficulties were to be encountered if he wished to put his property into mortmain. Hellouin was obliged to counterfeit madness, and at last to come to a very painful explanation with his liege lord; and, when he finally succeeded in obtaining the permission he craved, his establishment was so poor, that he was compelled to take upon himself the office of abbot, from an inability to find any other person who would accept it.—The monkish historians lavish their praises upon Hellouin. They assign to him every virtue under heaven; but they particularly laud him for his humility and industry: all day long he worked as a laborer in the building of his convent, whilst the night was passed in committing the psalter to memory. At this period of his life, a curious anecdote is recorded of him: curious in itself, as illustrative of the character of the man; and particularly curious, in being quoted as matter of commendation, and thus serving to illustrate the feelings of a great body of the community.—His mother, who shared in the pious disposition of her son, had attached herself to the convent to assist in the menial offices; and one day, while she was thus engaged, the building caught fire, and she perished in the flames; upon which, Hellouin, though bathed in tears, lifted up his hands to heaven, and gave thanks to God that his parent had been burned to death in the midst of an occupation of humility and piety!

During the life of Hellouin, the abbey was twice levelled with the ground: on each occasion it rose more splendid from its ruins, and on each the site was changed, till at length it was fixed upon the spot from which its ruins are now vanishing. The whole of Normandy would scarcely furnish a more desirable situation. Under the prelacy of Hellouin, Bec increased rapidly in celebrity, and consequently in the number of its inmates: it was principally indebted for this increase to an accidental circumstance. Lanfranc, a native of Pavia, a lawyer in Italy, but a monk in France, after having visited various monasteries, and distinguished himself by defending the doctrine of the real presence, then impugned by Berengarius, established himself here in the year 1042, and immediately opened a school, which, to judge from the language of Ordericus Vitalis[58], seems to have been the first ever known in Normandy. Scholars from France, from England, and from Flanders, hastened to place themselves under his care; his fame, according to William of Malmesbury, went forth into the outer parts of the earth; and Bec, under his auspices, became a most celebrated resort of literature. To borrow the more copious account given by William of Jumieges—"report quickly spread the glory of Bec, and of its abbot, Hellouin, through every land. The clergy, the sons of dukes, the most eminent schoolmasters, the most powerful of the laity, and the nobility, all hastened hither. Many, actuated by love for Lanfranc, gave their lands to the convent. The abbey was enriched with ornaments, with possessions, and with noble inmates. Religion and learning increased; property of all kinds abounded; and the monks, who but a few years before, could scarcely command sufficient ground for the site of their own building, now saw their estates extend for many miles in a lengthening line."—Promotion followed the fame of Lanfranc, who soon became abbot of the royal monastery of St. Stephen, at Caen, and thence was translated to the archiepiscopal see of Canterbury.

It was the rare good fortune of Bec, that the abbey furnished two successive metropolitans to the English church, both of them selected for their erudition, Lanfranc and Anselm. It is not a little remarkable, too, that both were Italians. Lanfranc, whilst archbishop of Canterbury, presided in the year 1077, at the dedication of the third church built at Bec. We may judge how far the abbey had at that time increased in consequence; for five bishops, one of them brother to the Conqueror, honored the ceremony with their presence; and the nobles and ladies of France, Normandy, and England crowded to the spot, to refresh their bodies by the pleasures of the festival, and their souls by endowments to the convent.

In the fifteenth century, when our Henry Vth brought his victorious armies into France, the monks of Bec were reduced to a painful alternative. It was apprehended by the French monarch, that the monastery might be converted into a dépôt by the English; and they were commanded either to demolish the church, or to fortify it against the invaders. They naturally regarded the latter as the lesser evil; and the consequence was, that the abbey was scarcely put into a state of defence, when it was attacked by the enemy, and, after sustaining a siege for a month, was obliged to surrender. A great part of the monastic buildings were levelled to the ground; and the fortifications which had been so strangely affixed to them were also razed: meanwhile the monks suffered grievously from the contending parties: their sacristy was plundered; their treasury emptied; and they were themselves exposed to a variety of personal hardships. At the same time, also, the tomb of the Empress Maud[59], which faced the high altar, was destroyed, after having been stripped of its silver ornaments.

Considering the number of illustrious persons who were abbots or patrons of Bec, and who had been elected from it to the superintendance of other monasteries, the church does not appear to have been rich in monuments. We read indeed of many individuals who were interred here belonging to the house of Neubourg, a family distinguished among the benefactors of the convent; and the records of the abbey speak also of the tomb of Richard of St. Leger, Bishop of Evreux; but the Empress was the only royal personage who selected this convent as the resting-place for her remains; and she likewise appears to have been the only eminent one, except Hellouin, the founder, who lay in the chapter-house, under a slab of black marble, with various figures of rude workmanship[60] carved upon it. His epitaph has more merit than the general class of monumental inscriptions:—

"Hunc spectans tumulum, titulo cognosce sepultum;

Est via virtutis nôsse quis ipse fuit.

Dum quater hic denos ævi venisset ad annos,

Quæ fuerant secli sprevit amore Dei.

Mutans ergò vices, mundi de milite miles

Fit Christi subito, Monachus ex laïco.
Hinc sibi, more patrum, socians collegia fratrum,
Curâ, quâ decuit, rexit eos, aluit.
Quot quantasque vides, hic solus condidit ædes,
Non tàm divitiis quàm fidei meritis.
Quas puer haud didicit scripturas postea scivit,
Doctus ut indoctum vix sequeretur eum.
Flentibus hunc nobis tulit inclementia mortis
Sextilis quinâ bisque die decimâ.
Herluine pater, sic cœlica scandis ovantèr;
Credere namque tuis hoc licet ex meritis."

In number of inmates, extent of possessions, and possibly, in magnificence of buildings, other Norman monasteries may have excelled Bec: none equalled it in the prouder honor of being a seminary for eminent men and especially for those destined to the highest stations in the church. Lanfranc and Anselm were not the only two of its monks who were seated on the archiepiscopal throne at Canterbury. Two others, Theobald and Hubert obtained the same dignity in the following century; and Roger, the seventh abbot of Bec, enjoyed the still more enviable distinction of having been unanimously elected to fill the office of metropolitan, but of possessing sufficient firmness of mind to resist the attractions of wealth, and rank, and power. The sees of Rochester, Beauvais, and Evreux were likewise filled by monks from Bec; and it was here that many monastic establishments, both Norman and foreign, found their pastors. Three of our own most celebrated convents, those of Chester, Ely, and St. Edmund's Bury, received at different epochs their abbots from Bec; and during the prelacy of Anselm, the supreme pontiff himself selected a monk of this house as the prior of the distant convent of the holy Savior at Capua.—The village of Bec, which adjoins the abbey, is small and unimportant.

I was returning to our carriage, when a soldier invited me to walk to a part of the monastic grounds (for they are very extensive) which is appropriated to the purpose of keeping up the true breed of Norman horses. The French government have several similar establishments: they consider the matter as one of national importance; and, as France has not yet produced a Duke of Bedford or a Mr. Coke, the state is obliged to undertake what would be much better effected by the energy of individuals.—A Norman horse is an excellent draft horse: he is strong, bony, and well proportioned. But the natives are not content with this qualified praise: they contend that he is equally unrivalled as a saddle-horse, as a hunter, and as a charger.

In this part of the country the present average price of a hussar's horse is nineteen pounds; of a dragoon's thirty-four pounds; and of an officer's eighty pounds.—These prices are considered high, but not extravagant. France abounds at this time in fine horses. The losses occasioned by the revolutionary wars, and more especially by the disastrous Russian campaign, have been more than compensated by five years of peace, and by the horses that were left by the allied troops. An annual supply is also drawn from Mecklenburg and the adjacent countries. Importations of this kind are regarded as indispensable, to prevent a degeneration in the stock. A Frenchman can scarcely be brought to believe it possible; that we in England can preserve our fine breed of horses without having recourse to similar expedients; and if at last, by dint of repeated asseverations, you succeed in obtaining a reluctant assent, the conversation is almost sure to end in a shrug of the shoulders, accompanied with the remark—"Ah, vous autres Anglais, vous voulez toujours voler de vos propres ailes."

As we approached Brionne, the face of the country became more uneven; and we passed an extensive tract of uncultivated chalk hills, resembling the downs of Wiltshire.—Brionne itself lies in a valley watered by the Risle: the situation is agreeable, and advantageous for trade. The present number of its inhabitants does not amount to two thousand; and there is no reason to apprehend that the population has materially decreased of late years. But in the times of Norman rule, Brionne was a town of more importance: it had then three churches, besides an abbey and a lazar-house. At present a single church only remains; and this is neither large, nor handsome, nor ancient, nor remarkable in any point of view. We found in it a monument of the revolution, which I never saw elsewhere, and which I never expected to see at all. The age of reason was a sadly irrational age.—The tablet containing the rights and duties of man, disposed in two columns, like the tables of the Mosaic law, is still suffered to exist in the church, though shorn of all its republican dignity, and degraded into the front of a pew.

On the summit of a hill that overhangs the town, stood formerly the castle of the Earls of Brionne; and a portion of the building, though it be but an insignificant fragment, is still left. The part now standing consists of little more than two sides of the square dungeon, The walls, which are about fifty feet in height, appear crumbling and ragged, as they have lost the greater part of their original facing. Yet their thickness, which even now exceeds twelve feet, may enable them to bid defiance for many a century, to "the heat of the sun, and the furious winter's rages."—Nearly the half of one of the sides, which is seventy feet long, is occupied by three flat Norman buttresses, of very small projection. No arched door-way, no window remains; nor any thing, except these buttresses, to give a distinct character to the architecture:

the hill is so overgrown with brush-wood, that though traces of foundation are discernible in almost every part of it, no clear idea can be formed of the dimensions or plan of the building. Its importance is sufficiently established by its having been the residence of a son or brother of Richard IInd, Duke of Normandy, on whose account, the town of Brionne, with the adjacent territory, was raised into an earldom. Historians speak unequivocally of its strength. During the reign of William the Conqueror, it was regarded as impregnable. This king was little accustomed to meet with disappointment or even with resistance; but the castle of Brionne defied his utmost efforts for three successive years. Under his less energetic successor, it was taken in a day. Its possessor, Robert, Earl of Brionne, felt himself so secure within his towers, that he ventured, with only six attendants, to oppose the whole army of the Norman Duke; but the besiegers observed that the fortress was roofed with wood; and a shower of burning missiles compelled the garrison to surrender at discretion.—The castle was finally dismantled by the orders of Charles Vth.

Brionne is known in ecclesiastical history as the place where the council of the church was held, by which the tenets of Berengarius were finally condemned. It appears that the archdeacon of Angers, after some fruitless attempts to make converts among the Norman monks, took the bold resolution of stating his doctrines to the duke in person; and that the prince, though scarcely arrived at years of manhood, acted with so much prudence on the occasion, as to withhold any decisive answer, till he had collected the clergy of the duchy. They assembled at Brionne, as a central spot; and here the question was argued at great length, till Berengarius himself, and a convert, whom he had brought with him, trusting in his eloquence, were so overpowered by the arguments of their adversaries, that they were obliged to renounce their errors. The doctrine of the real presence in the sacrament, was thus incontrovertibly established; and it has from that time remained an undisputed article of faith in the Roman Catholic church.

Footnotes:

[54] Vol. III. p. 187.—The engraving in the Antiquarian Repertory was made from a drawing in the possession of the late Sir William Burrell, Bart.

[55] The word Turold, in the tapestry, stands immediately over the head of a dwarf, who is holding a couple of horses; and it has therefore been inferred by Montfaucon, (Monumens de la Monarchie Française, I. p. 378.) that he is the person thus denominated. But M. Lancelot, in the Mémoires de

l'Académie des Inscriptions, VI. p. 753, supposes Turold to be the ambassador who is in the act of speaking; and this seems the more probable conjecture. The same opinion is still more decidedly maintained by Father Du Plessis, in his Histoire de la Haute Normandie, II. p. 342.—"Sur une ancienne tapisserie de l'Eglise de Baieux, que l'on croit avoir été faite par ordre de la Reine Mathilde femme du Conquérant, pour représenter les circonstances principales de cette mémorable expédition, on lit distinctement le mot Turold à côté d'un des Ambassadeurs, que Guillaume avoit envoiez au Comte de Ponthieu; et je ne doute nullement que ce Turold ne soit le même que le Connétable. Le sçavant Auteur des Antiquitez de notre Monarchie croit cependant que ce mot doit se rapporter à un Nain qui tient deux chevaux en bride derriere les Ambassadeurs; et il ajoute que ce Nain devoit être fort connu à la Conr du Duc de Normandie. On avoue que si c'est lui en effet qui doit s'appeller Turold, il devoit tenir aussi à la Cour de son Prince un rang distingué; sans quoi on n'auroit pas pris la peine de le désigner par son nom dans la tapisserie. On avoue encore que le nom de Turold est placé là de maniere qu'on peut à la rigueur le donner au Nain aussi bien qu'à l'un des deux Ambassadeurs; et comme le Nain est appliqué à tenir deux chevaux en bride, on pourrait croire enfin que c'est le Connétable, dont les titres de l'Abbaïe de Facan nous ont appris le nom: Signum Turoldi Constabularii. Mais le Nain est très-mal habillé, il a son bonnet sur la tête, et tourne le dos au Comte de Ponthieu, pendant que les deux Ambassadeurs noblement vêtus regardent ce Prince en face, et lui parlent découverts: trois circonstances qui ne peuvent convenir, ni au Connétable du Duc, ni à toute autre personne de distinction qui auroit tenu compagnie, ou fait cortege aux Ambassadeurs."

[56] This tower is figured, but very inaccurately, by Gough, in his Alien Priories, I. p. 22.—The cupola which then surmounted it is now gone; and the cap to the turret, which served as the staircase, has strangely changed its shape.

[57] Alien Priories, I. p. 24.

[58] "Nam antea, sub tempore sex ducum vix ullus Normannorum liberalibus studiis adhæsit; nec doctor inveniebatur, donec provisor omnium, Deus, Normannicis oris Lanfrancum appulit. Fama peritiæ illius in totâ

ubertim innotuit Europâ, unde ad magisterium ejus multi convenerunt de Franciâ, de Wasconiâ, de Britanniâ, necne Flandriâ."—Duchesne, Scriptores Normanni, p. 519.

[59] A question always existed, whether the Empress was really buried here, or at the abbey of Ste Marie des Prés, at Rouen. Hoveden expressly says, that she was interred at Rouen: the chronicle of Bec, on the other hand, is equally positive in the assertion that her body was brought to Bec, and entombed with honor before the altar of the Virgin. The same chronicle adds that, in the year 1273, her remains were discovered before the high altar, sewed up in an ox's hide.—Still farther to substantiate their claim, the monks of Bec maintained that, in 1684, upon the occasion of some repairs being done to this altar, the bones of the empress were again found immediately under the lamp (which, in Catholic churches, is kept constantly burning before the holy sacrament,) and that they were deposited once more in the ground in a wooden chest, covered with lead.—The Empress was a munificent endower of monasteries, and was at all times most liberal towards Bec. William of Jumieges says, that it would be tedious to enumerate the presents she made to the abbey, but that the sight of them gave pleasure to those strangers who have seen the treasures of the most noble churches. His remarks on this matter, and his account of her arguments with her father, on the subject of her choice of Bec, as a place of her interment, deserve to be transcribed.—"Transiret illac hospes Græcus aut Arabs, voluptate traheretur eadem. Credimus autem, et credere fas est, æquissimum judicem omnium non solùm in futuro, verumetiam in præsenti seculo, illi centuplum redditurum, quod seruis suis manu sicut larga, ita devota gratantèr impendit. Ad remunerationem verò instantis temporis pertinere non dubium est, quòd, miserante Deo, sopita adversa valetudine, sanctitatem refouit, et Monachos suos, Monachos Beccenses, qui præ omnibus, et super omnes pro ipsius sospitate, jugi labore supplicandi decertando pene defecerant, aura prosperæ valetudinis ejus afflatos omninò redintegravit.—Nec supprimendum illud est silentio, imò, ut ita dicatur, uncialibus literis exaratum, seculo venturo transmittendum; quòd antequam convalesceret postulaverat patrem suum, ut permitteret eam in Cœnobio Beccensi

humari. Quod Rex primo abnuerat, dicens non esse dignum, ut filia sua, Imperatrix Augusta, quæ semel et iterùm in urbe Romulea, quæ caput est mundi, per manus summi Pontificis Imperiali diademate processerat insignita, in aliquo Monasterio, licèt percelebri et religione et fama, sepeliretur; sed ad civitatem Rotomagensium, quæ metropolis est Normannorum, saltem delata, in Ecclesia principali, in qua et majores ejus, Rollonem loquor et Willelmum Longamspatam filium ipsius, qui Neustriam armis subegerunt, positi sunt, ipsa et poneretur. Qua deliberatione Regis percepta, illi per nuncium remandavit, animam suam nunquam fore lætam, nisi compos voluntatis suæ in hac duntaxat parte efficeretur.—O femina macte virtutis et consilii sanioris, paruipendens pompam secularem in corporis depositione! Noverat enim salubrius esse animabus defunctorum ibi corpora sua tumulari, ubi frequentiùs et devotiùs supplicationes pro ipsis Deo offeruntur. Victus itaque pater ipsius Augustæ pietate et prudentia filiæ, qui ceteros et virtute et pietate vincere solitus erat, cessit, et voluntatem, et petitionem ipsius de se sepelienda Becci fieri concessit. Sed volente Deo ut præfixum est, sanitati integerrimæ restituta convaluit."—Duchesne, Scriptores Normanni, p. 305.

[60] Histoire de la Haute Normandie, II. p, 281.

LETTER XXI
BERNAT—BROGLIE—ORBEC—LISIEUX—CATHEDRAL—ECCLESIASTICAL HISTORY

(*Lisieux, July,* 1818.)

Instead of pursuing the straight road from Brionne to this city, we deviated somewhat to the south, by the advice of M. Le Prevost; and we have not regretted the deviation.

Bernay was once celebrated for its abbey, founded in the beginning of the eleventh century, by Judith, wife of Richard IInd, Duke of Normandy. Some of the monastic buildings are standing, and are now inhabited: they appear to have been erected but a short time before the revolution, and to have suffered little injury.—But the abbey church, which belonged to the original structure, is all desolate within, and all defaced without. The interior is divided into two stories, the lower of which is used as a corn market, the upper as a cloth hall. Thus blocked up and encumbered, we may yet discern that it is a noble building: its dimensions are grand, and in most parts it is a perfect specimen of the semi-circular style, except the windows and the apsis, which are of later dates. The pillars in the nave and choir are lofty, but massy: the capitals of some of them are curiously sculptured. On the lower member of the entablature of one capital there are still traces of an inscription; but it is so injured by neglect and violence, that we were unable to decipher a single word. The capital itself is fanciful and not devoid of elegance.

The convent was placed under the immediate protection of the sovereign, by virtue of an ordinance issued by Philip Augustus[61], in 1280, at which time Peter, Count of Alençon, attempted to establish a claim to some rights affecting the monastery. He alleged a grant from a former monarch to one of his predecessors, by whom he asserted that the convent had been founded; and, in support of his claim, he urged its position within the limits of his territory. The abbot and monks resisted: they gave proof that the abbey of Bernay was really founded by the duchess; and therefore the king, after a full and impartial hearing, decided against the count, and declared that the advocation of the monastery was thenceforth to belong to himself and his successors in the dukedom for ever.—Judith died before the convent was entirely built, and the task of completing it devolved upon her widowed husband, whose charter, confirming the foundation, is still in existence. It begins by a recital of the pious motives[62] which urged the duchess to the undertaking; it expressly mentions her death while the building was yet unfinished; and, after detailing the various lands and grants bestowed on the abbey, it concludes by denouncing the anger of God, and a fine of two hundred pounds weight of gold upon those who disturb the establishment, "that they may learn to their confusion that the good deeds of their ancestors, undertaken for the love of God, are not to be undone with impunity."

The parochial church at Bernay is uninteresting. The sculptures, however, which adorn the high altar, are relics saved from the destruction of the abbey of Bec. The Virgin Mary and Joseph are represented, contemplating the infant Jesus, who is asleep. The statues are all of the natural size. We saw many grave-stones from the same abbey, nine or ten feet long, and covered with monumental figures of the usual description, indented in the stone. These memorials were standing by the side of the church door, not for preservation, but for sale! And at a small chapel in the burial-ground near the town, we were shewn twelve statues of saints, which likewise came from Bec. They are of comparatively modern workmanship, larger than life, and carved in a good, though not a fine, style. In the same chapel is kept the common coffin for the interment of all the poor at Bernay.

The custom of merely putting the bodies of persons of the lower class into coffins, when they are brought to the burial-ground, and then depositing them naked in their graves, prevails at present in this part of France as it did formerly in England.—In a place which must be the receptacle for many that were in easy, and for not a few that were in affluent, circumstances, it was remarkable that all lay indiscriminately side by side, unmarked by any monumental stone, or any sepulchral record.—Republican France proscribed distinctions of every description, and those memorials which tended to perpetuate distinctions beyond the limits of mortal existence, were naturally most unpardonable in the eyes of the apostles of equality. But doctrines of this nature have fallen into disrepute for more than twenty years; and yet the country church-yard remains as naked as when the guillotine would have been the reward of opposition to the tenets of the day. There are few more comfortless sights, than such a cemetery: it looks as if those by whom it is occupied regarded death as eternal sleep, and thought that the memory of man should terminate with the close of his life. However unlettered the muse, however hackneyed the rhyme, however misapplied the text, it is consolatory to see them employed. Man dwells with a melancholy satisfaction upon the tomb-stones of his relations and friends, and not of them alone, but of all whom he has known or of whom he has heard.—A mere hic jacet, with the name and years of him that sleeps beneath, frequently recals the most lively impressions; and he who would destroy epitaphs would destroy a great incitement to virtue.—In other parts of France tomb-stones, or crosses charged with monumental inscriptions, have re-appeared: at Bernay we saw only two; one of them commemorated a priest of the town; the other was erected at the public expence, to the memory of three gendarmes, who were killed at the beginning of the revolution, and before religion was proscribed, in the suppression of some tumult.

At less than a mile from Bernay, in the opposite direction, is another church, called Notre Dame de la Couture, a name borrowed from the property on which it stands. We were induced to visit it, by the representation of different persons in the town, who had noticed our architectural propensities. Some assured us that "C'est une belle pièce;" others that "C'est une pièce qui n'est pas vilaine;" and all concurred in praising it, though some only for the reason that "les processions vont tout autour du choeur."—We found nothing to repay the trouble of the walk.

Bernay contains upwards of six thousand inhabitants, the greater part of whom are engaged in manufacturing coarse woollen and cotton cloths; and the manufactures flourish, the goods made being principally for home consumption. It is the chief place of the arrondissement, and the residence of a sub-prefect.—Most of the houses are like those at Rouen, merely wooden frames filled with mortar, which, in several instances, is faced with small bricks and flints, disposed in fanciful patterns: here and there the beams are carved with a variety of grotesque figures. The lower story of all those in the high street retires, leaving room for a wooden colonnade, which shelters the passenger, though it is entirely destitute of all architectural beauty. The head-dress of the females at Bernay is peculiar, and so very archaic, that our chamber-maid at the inn appeared to deserve a sketch, full as much as any monumental effigy.

On our road between Bernay and Orbec, we stopped at the village of Chambrais, more commonly called Broglie. Before the revolution, it belonged to the noble family of that name, and it thence derived its familiar appellation. The former residence of the Seigneurs of Broglie, which is still standing, apparently uninjured, upon an adjoining eminence, has lately been restored to the present Maréchal Duc de Broglie. It looks like an extensive parish work-house, or like any thing rather than a nobleman's seat.—The village church is very ancient and still curious, though in parts considerably modernized. Unlike most churches of great antiquity, it is not built in the form of a cross, but consists only of a nave and choir, with side-aisles and an apsis, all on a small scale[63]. Towards the north, the nave is separated from the aisle by some of the largest and rudest piers I ever saw. They occupy full two-thirds of the width of the intervening arches, which are five feet wide, elliptic rather than semi-circular, and altogether without ornament of any kind. Above each of these arches is a narrow, circular-headed window, banded with a cylindrical pilaster; and, in most instances, a row of quatrefoils runs between the pillar and the window. The bases of the windows rest upon a string-course that extends round the whole building; and on this also, alternating with the windows, rest corbels, from which spring very short, clustered columns, intended to support the groinings of the roof. On the south side, the massy piers have been pared into comparatively slender pillars; and the arches are pointed, as are all the lower windows in the church.—The font is of stone, and ancient: it consists of a round basin, on a quadrangular pedestal, like many in England.—The west front of the church is peculiar. It is entered by a very wide, low, semi-circular door-way, of rude architecture, and quite unornamented. Above is a window corresponding with those in the clerestory; and, still higher, a row of interlaced arches, also semi-circular. A pointed arch, the receptacle for the statue of a saint, surmounts the whole; but this is, most probably, of a later æra, as evidently are the two lateral compartments, which terminate in slender spires of slate, and are separated from the central division by Norman buttresses.

We stopped to dine at Orbec, a small and insignificant country town, formerly an appendage of the houses of Orléans and Navarre, with the title of a barony; but, more immediately before the revolution, the domain of the family of Chaumont. Its church is a most uncouth edifice: the plan is unusual; the entrance is in the north transept, which ends in a square high tower.

Bernay, Orbec, and Lisieux, communicate only by cross roads, scarcely passable by a carriage, even at this season of the year. From Orbec to Lisieux the road runs by the side of the Touques, which, at Orbec, is no more than a rivulet. The beautiful green meadows in the valley, appear to repay the great care which is taken in the draining and irrigating of them. They are every where intersected by small trenches, in which the water is confined by means of sluices.—In this part of the country, we passed several flocks of sheep, the true moutons du pays, a large breed, with red legs and red spotted faces. Their coarse wool serves to make the ordinary cloth of the country, but is inapplicable to any of a finer texture. To remedy this deficiency, and, if possible, improve the local manufactures, some large flocks of Merino sheep were imported at the time when the French occupied Spain; and they are said to thrive. But it is only of late years that any attempts, have been made of the kind.—The Norman farmer, however careful about the breed of his horses, has altogether neglected his sheep; and this is the more extraordinary, considering that the prosperity of the province is inseparably connected with that of the manufactures, and that much of the value of the produce must of necessity depend upon the excellence of the material. His pigs are the very perfection of ugliness: it is no hyperbole to say, that, in their form, they partake as much of a greyhound as of an English pig.—These animals are sure to attract the gaze of our countrymen; and poor Trotter, in his narrative of the journey of Mr. Fox, expressed his marvel so often, as to call down upon himself the witty vengeance of one of our ablest periodical writers.

Melons are cultivated on a great scale in the country about Lisieux. They grow here in the natural soil, occupying whole fields of considerable size, and apparently without requiring any extraordinary pains.—As we approached the city, the meadows, through which we passed, were mostly occupied as extensive bleaching-grounds. Lisieux is an industrious manufacturing town. Its ten thousand inhabitants find their chief employment in the making of the ordinary woollen cloths, worn by the peasantry of Normandy and of Lower Brittany. Linen and flannels are also manufactured here, though on a comparatively trifling scale. For trade of this description, Lisieux is well situated upon the banks of the Touques, a small river, which, almost immediately under the walls of the town, receives the waters of a yet smaller stream, the Orbec. A project is in agitation, and it is said that it may be carried into effect at an inconsiderable expence, of making the Touques navigable to Lisieux. At present, it is so no farther than the the little town of the same name as the river; and even this derives no great advantage from the navigation; for, however near its situation is to the mouth of the stream, it is approachable only by vessels of less than one hundred tons burthen.—It was at Touques that Henry Vth landed in France, in the spring of 1417, when the monarch, flushed with a degree of success as extraordinary as it was unexpected, quitted England with the determination of returning no more till the whole kingdom of France should be subjugated.

The greater part of the houses in Lisieux are built of wood; and many of them are old, and most of them are mean; yet, on the whole, it is picturesque and handsome. Its streets are spacious, and contain several large buildings: it is surrounded with pleasant boulevards; and its situation, like that of most other Norman towns, is delightful.—In consequence of the revolution, the city has lost the privilege of being an episcopal see. Even when Napoléon, by virtue of the concordat of 1801, restored the Gallican church to its obedience to the the supreme Pontiff, the see of Lisieux was suppressed. The six suffragan bishops of ancient Normandy were at that time reduced to four, conformably to the number of the departments of the province; and Lisieux and Avranches merged in the more important dioceses of Bayeux and Coutances.

The cathedral, now the parish church of St. Peter, derived, however, one advantage from the revolution. Another church, dedicated to St. Germain, which had previously stood immediately before it, so as almost to block up the approach, was taken down, and the west front of the cathedral was made to open upon a spacious square.—Solid, simple grandeur are the characters of this front, which, notwithstanding some slight anomalies, is, upon the whole, a noble specimen of early pointed architecture.—It is divided into three equal compartments, the lateral ones rising into short square towers of similar height. The southern tower is surmounted by a lofty stone spire, probably of a date posterior to the part below. The spire of the opposite tower fell in 1553, at which time much injury was done to the building, and particularly to the central door-way, which, even to the present day, has never been repaired.—Contrary to the usual elevation of French cathedrals, the great window over the principal entrance is not circular, but pointed: it is divided into three compartments by broad mullions, enriched with many mouldings. The compartments end in acute pointed arches.—In the north tower, the whole of the space from the basement story is occupied by only two tiers of windows. Each tier contains two windows, extremely narrow, considering their height; and yet, narrow as they are, each of them is parted by a circular mullion or central pillar. You will better understand how high they must be, when told that, in the southern tower, the space of the upper row is divided into three distinct tiers; and still the windows do not appear disproportionately short. They also are double, and the interior arches are pointed; but the arches, within which they are placed, are circular. In this circumstance lies the principal anomaly in the front of the cathedral; but there is no appearance of any disparity in point of dates; for the circular arches are supported on the same slender mullions, with rude foliaged capitals, of great projection, which are the most distinguishing characteristics of this style of architecture.

The date of the building establishes the fact of the pointed arch being in use, not only as an occasional variation, but in the entire construction of churches upon a grand scale, as early as the eleventh century.—Sammarthanus tells us that Bishop Herbert, who died in 1049, began to build this church, but did not live to see it completed; and Ordericus Vitalis expressly adds, that Hugh, the successor to Herbert, upon his death-bed, in 1077, while retracing his past life, made use of these words:—"Ecclesiam Sancti Petri, principis apostolorum, quam venerabilis Herbertus, praedecessor meus, coepit, perfeci, studiosè adornavi, honorificè dedicavi, et cultoribus necessariisque divino servitio vasis aliisque apparatibus copiosè ditavi."—Language of this kind appears too explicit to leave room for ambiguity, but an opinion has still prevailed, founded probably upon the style of the architecture, that the cathedral was not finished till near the expiration of the thirteenth century. Admitting, however, such to be the fact, I do not see how it will materially help those who favor the opinion; for the building is far from being, as commonly happens in great churches, a medley of incongruous parts; but it is upon one fixed plan; and, as it was begun, so it was ended.—The exterior of the extremity of the south transept is a still more complete example of the early pointed style than the west front: this style, which was the most chaste, and, if I may be allowed to use the expression, the most severe of all, scarcely any where displays itself to greater advantage. The central window is composed of five lancet divisions, supported upon slender pillars: massy buttresses of several splays bound it on either side.

The same character of uniformity extends over the interior of the building. On each side of the nave is a side-aisle; and, beyond the aisles, chapels. The pillars of the nave are cylindrical, solid, and plain. Their bases end with foliage at each corner, and foliage is also sculptured upon the capitals. The arches which they support are acute.—The triforium is similar in plan to the part below; but the capitals of the columns are considerably more enriched, with an obvious imitation of the antique model, and every arch encircles two smaller ones. In the clerestory the windows are modern.—The transepts appear the oldest parts of the cathedral, as is not unfrequently the case; whether they were really built before the rest, or that, from being less used in the services of the church, they were less commonly the objects of subsequent alterations. They are large; and each of them has an aisle on the eastern side. The architecture of the choir resembles that of the nave, except that the five pillars, which form the apsis, are slender and the intervening arches more narrow and more acute.—The Lady-Chapel, which is long and narrow, was built towards the middle of the fifteenth century, by Peter Cauchon, thirty-sixth bishop of Lisieux, who, for his steady attachment to the Anglo-Norman cause, was translated to this see, in 1429, when Beauvais, of which he had previously been bishop, fell into the hands of the French. He was selected, in 1431, for the invidious office of presiding at the trial of the Maid of Orléans. Repentance followed; and, as an atonement for his unrighteous conduct, according to Ducarel, he erected this chapel, and therein founded a high mass to the Holy Virgin, which was duly sung by the choristers, in order, as is expressed in his endowment-charter, to expiate the false judgment which he pronounced[64].—The two windows by the side of the altar in this chapel have been painted of a crimson color, to add to the effect produced upon entering the church; and, seen as they are, through the long perspective of the nave and the distant arches of the choir, the glowing tint is by no means unpleasing.—The central tower is open within the church to a considerable height: it is supported by four arches of unusual boldness, above which runs a row of small arches, of the same character as the rest of the building; and, still higher, on each side, are two lancet-windows.—The vaulting of the roof is very plain, with bosses slightly pendant and carved.

At the extremity of the north transept is an ancient stone sarcophagus, so built into the wall, that it appears to have been incorporated with the edifice, at the period when it was raised. The style of the medallions which adorn it will be best understood by consulting the annexed sketch, which is very faithful, though taken under every possible disadvantage. The transept is now used as a school; and the little filthy imps, who are there taught to drawl out their catechisms, continued swarming round the feverish artist, during the progress of the drawing. The character of the heads, the crowns, and the disposition of the foliage, may be considered as indicating that it is a production, at least of the Carlovingian period, if it be not indeed of earlier date. I believe it is traditionally supposed to have been the tomb of a saint, perhaps St. Candidus; but I am not quite certain whether I am accurate in the recollection of the name.—Above are two armed statues, probably of the twelfth or thirteenth centuries. These have been engraved by Willemin, in his useful work, Les Monumens Français, under the title of Two Armed Warriors, in the Nave of the Cathedral at Lisieux; and both are there figured as if in all respects perfect, and with a great many details which do not exist, and never could have existed, though at the same time the draftsman has omitted the animals at the feet of the statues, one of which is yet nearly entire.—This may be reckoned among the innumerable proofs of the disregard of accuracy which pervades the works of French antiquaries. A

French designer never scruples to sacrifice accuracy to what he considers effect.—Willemin describes the monuments as being in the nave of the church. I suspect that he has availed himself of the unpublished collection of Gaignat, in this and many other instances. It is evident that originally the statues were recumbent; but I cannot ascertain when they changed their position.—No other tombs now exist in the cathedral: the brazen monument raised to Hannuier, an Englishman, the marble that commemorated the bishop, William d'Estouteville, founder of the Collège de Lisieux at Paris, that of Peter Cauchon in the Lady-Chapel, and all the rest, were destroyed during the revolution.

The diocese of Lisieux was a more modern establishment than any other in Normandy. Even those who are most desirous to honor it by antiquity, do not venture to date its foundation higher than the middle of the sixth century. Ordericus Vitalis, a monk of the province, suggests with some reason that we ought not to be hasty in forming our judgment upon these subjects; for that, owing to the destruction caused by the Norman pirates and the abominable negligence (*damnabilis negligentia*) of those to whom the care of the records of religious houses had subsequently been intrusted, many documents had been irretrievably lost.—The see of Lisieux was also peculiarly unfortunate, in having twice been in a state of anarchy, and on each occasion for a period of more than a century. The series of its prelates is interrupted from the year 670 to 853, and again from 876 to 990.

It is rather extraordinary, that no one of the Lexovian bishops was ever admitted by the church into the catalogue of her saints. Many of them were prelates of unquestionable merit. Freculfus, in the ninth century, was a patron of literature, and himself an author; Hugh of Eu, grandson of Richard, Duke of Normandy, was one of the most illustrious ecclesiastics of his day; Gilbert is described by Ordericus Vitalis as having been a man of exemplary charity, and deeply versed in all sciences, though it is admitted that he was somewhat too much addicted to worldly pleasures, and not averse from gambling; and Arnulf, whose letters and epigrams are preserved among the manuscripts of the Vatican, was a prelate who would have done honor to St. Peter's chair.—All these were bishops of Lisieux, during the ages when canonization was not altogether so unfrequent as in our days. Arnulf particularly distinguished himself by taking a leading part in the principal transactions of the times. He accompanied the crusaders to the holy land in 1147; five years subsequently he officiated at the marriage of Henry Plantagenet with Eleanor of Guyenne, the repudiated wife of Louis le Jeune, which was performed in his cathedral; he assisted at the coronation of the same king, by whom he was shortly afterwards employed in a mission of great importance at Rome; and he interposed to settle the differences between that sovereign and Thomas à Becket; and though he espoused the part of the prelate, he had the good fortune to retain the favor of the monarch. A life thus eventful ended with the conviction that all was vanity!--Arnulf, disgusted with sublunary honors, abdicated his see and retired to a monastery at Paris, where he died.—One of the immediate successors of this prelate, William of Rupierre, was the ambassador of Richard Coeur-de-Lion to the Pope; and he pleaded the cause of his sovereign against Walter, Archbishop of Rouen, on the occasion of the differences that originated from the building of Château Gaillard. He also resisted the power usurped by King John within the city and liberties of Lisieux, and finally obtained a sentence from the Norman court of exchequer, whereby the privileges of the dukes of the province were restricted to what was called the Placitum Spathæ, consisting of the right of billetting soldiers, of coining money, and of hearing and determining in cases of appeal. The decision is honorable both to the independence of the court, and the vigor of the prelate.—In times nearer to our own, a bishop of Lisieux, Jean Hennuyer, obtained a very different distinction. Authors are strangely at variance whether this prelate is to be regarded as the protector or the persecutor of the protestants. All agree that his church suffered materially from the excesses of the Huguenots, in 1562, and that, on the following year, he received public thanks from the Cardinal of Bourbon, for the firmness with which he had opposed them; but the point at issue is, whether, after the massacre of St. Bartholomew, ten years subsequently, he withstood the sanguinary orders from the court to put the Huguenots to the sword, or whether he endeavored, as far as lay in his power, to forward the pious labor of extirpating the heretics, but was himself effectually resisted by the king's own lieutenant.—Sammarthanus tells us that the first of these traditions rests solely upon the authority of Anthony Mallet[65] but it obtained general credence till within the last three years, when a very well-informed writer, in the Mercure de France, and subsequently in the article Hennuyer in the Bibliographie Universelle, espoused, and has apparently established, the opposite opinion.

We visited only one other of the churches in Lisieux, that of St. Jacques, a large edifice, in a bad style of pointed architecture, and full of gaudy altars and ordinary pictures. On the outside of the stalls of the choir towards the north is some curious carving; but I should scarcely have been induced to have spoken of the building, were it not for one of the paintings, which, however uninteresting as a piece of art, appears to possess some historical value. It represents how the bones of St. Ursinus were miraculously translated to Lisieux, under the auspices of Hugh the Bishop, in 1055; and it professes, and apparently with truth, to be a copy, made in the seventeenth century, from an original of great antiquity. The legend relating to the relics of this saint, is noticed by no author with whom I am acquainted, nor do I find him mentioned any where in conjunction with the church of Lisieux, or with any other Norman diocese.—But the extraordinary privilege granted to the canons of the cathedral, of being Earls of Lisieux, and of exercising all civil and criminal jurisdiction within the earldom, upon the vigil and feast-day of St. Ursinus, in every year, is most probably connected with the tradition commemorated by the picture. The actual existence of the privilege, in modern times, we learn from Ducarel; who also details at length the curious ceremonies with which the claim of it was accompanied. The exercise of these rights was confirmed by a compact between the canons and the bishop, who, prior to the revolution, united the secular coronet of an earl with the episcopal mitre, and bore supreme sway in all civil and ecclesiastical polity, during the remaining three hundred and sixty-three days in the year.

Footnotes:

[61] This ordinance is preserved by Du Monstier in the Neustria Pia, p. 400.

[62] The preamble of the charter is as follows:—"Nulli dubium videri debet futuros esse haeredes Regni coelestis, et cohaeredes Dei, qui Christum haeredem sui facientes, eorum, quæ in hujus vitae peregrinatione, quasi a quadam paterna haereditate possident, locis ea Divino cultui deditis mancipare non dubitant. Ad quam rem, nostram firmat fidem calix aquæ frigidae, qui, juxta Evangelicum verbum, suo pollet munere. Non ergò divini muneris gratia privari credendi sunt, qui Ecclesiasticis obsequiis, etsi officio non intersunt, rerum tamen suarum admistratione, Divini officii sustentant ministros: ea spe temporalem subministrantes alimoniam, ut sic solummodò coelestibus reddant intentos, qui coelestis Regis assiduo constituuntur invigilare obsequio, participes fiant ejusmodi beneficii omnimodò."—Neustria Pia, p. 398.

[63] The following are the dimensions of the building, in English feet:—

	LENGTH.	WIDTH.
Nave	54	15
Choir	45	15
North aisle	7	
South ditto	15	

[64] Anglo-Norman Antiquities, p. 47.

[65] "Sed ne quid omittam eorum etiam quæ unum Antonium Mallet habent auctorem, anno 1572, cum prorex urbis Lexoviensis Livarotus a Carolo rege literas accepisset, quibus qui Lexovii infecti erant hæresi occidi omnes jubebantur per eos dies quibus princeps civitas cruore ejus insaniæ hominum commaduerat, easque communicasset episcopo: Neque sum passurus, inquit præsul, oves meas, et quamquam evagatas Christi caula, meas tamen adhuc, necdum desperatas, gladio trucidari. Referente contra prorege imperio se mandatoque urgeri principis; quod si posthabeatur, omnem esse periculi aleam in caput suum moriendique necessitatem redituram: Et polliceor, inquit episcopus, illa te eximendum, postulantique cautionem, præsul consignatum manu sua scriptum tradidit, fidem datam confirmans. Qua illico publicata clementia, et ad errantes oves perlata, sollicitudine præsulis vigilantis circa gregis commissi sibi salutem et conservationem, rediere sensim in ecclesiæ sinum omnes quotquot Lexovii per ea tempora novum istud fataleque delirium dementarat, nec ultra ibidem diu visi qui a recta fide aberrarent."—Gallia Christiana, p. 802.

LETTER XXII
SITE AND RUINS OF THE CAPITAL OF THE LEXOVII—HISTORY OF LISIEUX—MONASTERIES OF THE DIOCESE—ORDERICUS VITALIS—M. DUBOIS—LETTER FROM THE PRINCESS BORGHESE

(*Lisieux*, *July*, 1818.)

Lisieux represents one of the most ancient capitals of the primitive tribes of Gaul. The Lexovii, noticed by Julius Cæsar, in his *Commentaries*, and by other authors, who were almost contemporary with the Roman conqueror, are supposed by modern geographers to have occupied a territory nearly co-extensive with the bishopric of Lisieux; and it may be remarked, that the bounds of the ancient bishoprics of France were usually conterminal with the Roman provinces and prefectures.

The capital of the Lexovii was called the Neomagus or Noviomagus Lexoviorum; and no doubt ever was entertained but that the present city occupied the same site, till an accidental discovery, in the year 1770, proved the contrary to be the fact.—About that time a chaussée was formed between Lisieux and Caen; and, in the course of some excavations, which were made under the direction of M. Hubert, the superintending engineer, for the purpose of procuring stone, the laborers opened the foundations of some ruined buildings scattered over a field, called les Tourettes, about three-quarters of a mile from the former town. The character of these foundations was of a nature to excite curiosity: they were clearly the work of a remote age, and various specimens of ancient art were dug up amongst the ruins. The extent of the foundations, which spread over a space four times as large as the plot occupied by modern Lisieux left no doubt but that Danville, and all other geographers, must have been mistaken with respect to the position assigned by them to the ancient Neomagus. M. Hubert drew a plan of the ruins, and accompanied it with an historical memoir; but unfortunately he was a man little capable of prosecuting such researches; and though M. Mongez, in his report to the National Institute[66], eulogized the map as exact, and the memoir as excellent, they were both of them extremely faulty.

It was reserved for M. Louis Dubois, of whom I shall have occasion to speak again before I close this letter, to repair the omissions and rectify the mistakes of M. Hubert, and he has done it with unremitting zeal and extraordinary success. The researches of this gentleman, among the remains of Neomagus Lexoviorum, have already brought to light a large number of valuable medals, both in silver and bronze, as well as a considerable quantity of fragments of foreign marble, granite, and porphyry, some of them curiously wrought. The most important of his discoveries has been recently made: it is that of a Roman amphitheatre, in a state of great perfection, the grades being covered only by a thin layer of soil, which a trifling expence of time and labor will effectually remove.

Such vestiges prove that Neomagus must have been a place of importance; and, like the other Gallo-Roman cities, it would probably have maintained its honors under the Franks; but about the middle of the fourth century, the Saxons, swarming from the mouths of the Elbe and Weser, laid waste the coasts of Belgium and of Neustria, and finally established themselves in that portion of northern Gaul called the Secunda Lugdunensis, which thence obtained, in the Notitia Imperii, the title of the Littus Saxonicum.—In the course of these incursions, it is supposed that Neomagus was utterly destroyed by the invaders. None of the medals dug up within the precincts of the town, or in its neighborhood, bear a later date than the reign of Constantine; and, though the city is recorded in the Itinerary of Antoninus, no mention of it is to be found in the curious chart, known by the name of the Tabula Peutingeriana, formed under the reign of Theodosius the Great; so that it then appears to have been completely swept away and forgotten.

The new town of Lisieux and the bishopric most probably arose together, towards the close of the sixth century; and the city, like other provincial capitals in Gaul, took the name of the tribe by whom the district had been peopled. It first appears in history under the appellation of Lexovium or Lexobium: in the eleventh century, when Ordericus Vitalis composed his history, it was called Luxovium; and soon after it became Lixovium, and Lizovium, which, gallicised, naturally passed into Lyzieulx, or, as it is now written, Lisieux. The city was ravaged by the Normans about the year 877, in the course of one of their predatory excursions from Bayeux: it again felt their vengeance early in the following century, when Rollo, after taking Bayeux by storm, sacked Lisieux at the head of his army on his way to Rouen. The conqueror was not put in possession of the Lexovian territory by Charles the Simple till 923, eleven years after the rest of Neustria had been ceded to him.

United to the duchy, Lisieux enjoyed a short respite from the calamities of war; nor does it appear to have borne any prominent part in the transactions of the times. The name, indeed, of the city occurs as the seat of the council held for the purpose of degrading Malgerius from the primacy of Normandy; but, except on this occasion, Lisieux is scarcely mentioned till the first year of the twelfth century, when it was the seat of rebellion. Ralph Flambart, bishop of Durham, a prelate of unbounded arrogance, had fled from England, and joined Duke Robert, then in arms against his brother. Raising the standard of insurrection, he fixed himself at Lisieux, took forcible possession of the town, and invested his son, only twelve years old, with the mitre[67], while he himself exercised despotic authority over the inhabitants. At length, he purchased peace and forgiveness, by opening the gates to his lawful sovereign, after the battle of Tinchbray.—In the middle of October, in the same year, Henry returned to Lisieux, and there held an assembly of the Norman nobility and prelates, who proclaimed peace throughout the duchy, enacted sundry strict regulations to prevent any infringement of the laws, and decreed that Robert, the captive duke, should be consigned to an English prison.—Two years subsequently, another council was also assembled at Lisieux, by the same sovereign, and for nearly the same objects; and again, in 1119, Henry convened his nobles a third time at Lisieux, when this parliament ratified the peace concluded at Gisors, six years previously, and witnessed the marriage[68] of the king's son, William Adelin, with Matilda, daughter of Fulk, earl of Anjou.

Historical distinction is seldom enviable:—in the wars occasioned by the usurpation of Stephen, Lisieux once more obtained an unfortunate celebrity. The town was attacked in 1136, by the forces of Anjou, under the command of Geoffrey Plantagenet, husband of the Empress Maud, joined by those of William, Duke of Poitiers; and the garrison, consisting of Bretons, seeing no hope of effectual resistance or of rescue, set fire to the place to the extreme mortification of the invaders, who, in the language of the chronicles of the times, "when they beheld the city and all its wealth a prey to the flames, waxed exceedingly wroth, at being deprived of the spoil; and grieved sorely for the loss of the booty which perished in the conflagration."—The town, however, was not so effectually ruined, but that, during the following year, it served King Stephen as a rallying point, at which to collect his army to march against his antagonist.—In 1169, it was distinguished by being selected by Thomas à Becket, as the place of his retirement during his temporary disgrace.

History from this time forward relates but little concerning Lisieux. Though surrounded with walls during the bishopric of John, who was promoted to the see early in the twelfth century, the situation of the town, far from the coast or from the frontiers of the province, rendered the inhabitants naturally unwarlike, and caused them in general to submit quietly to the stronger party.—Brito, in his Philippiad, says that, when Philip Augustus took Lisieux, in 1213, the Lexovians, destitute of fountains, disputed with the toads for the water of the muddy ditches. His mentioning such a fact is curious, as shewing that public fountains were at that early period of frequent occurrence in Normandy.—Our countrymen, in the fifteenth century, acted with great rigor, to use the mildest terms, towards Lisieux. Henry, after landing at Touques, in 1417, entered the town, in the character of an enraged enemy, not as the sovereign of his people: he gave it up to plunder; and even the public archives were not spared. The cruelty of our English king is strongly contrasted by the conduct of the Count de Danois, general of the army of Charles VIIth, to whom the town capitulated in 1449. Thomas Basin, then bishop, negociated with such ability, that, according to Monstrelet, "not the slightest damage was done to any individual, but each peaceably enjoyed his property as before the surrender."

The most celebrated monasteries within the diocese of Lisieux were the Benedictine abbeys of Bernay, St. Evroul, Preaux, and Cormeilles.—Cormeilles was founded by William Fitz-Osborne, a relation to William the Conqueror, at whose court he held the office of sewer, and by whom he was promoted to the earldom of Hereford. Its church and monastic buildings had so far gone to ruin, in the last century, as to call forth a strong remonstrance from Mabillon[69]: they were afterwards repaired by Charles of Orléans, who was appointed abbot in 1726.—The abbey of Preaux is said to have existed prior to the invasion of the Normans; but its earliest records go no farther back than the middle of the eleventh century, when it was restored by Humphrey de Vetulis, who built and inclosed the monastery about the year 1035, at which time Duke Robert undertook his pilgrimage to the Holy Land. This abbey, according to the account given by Gough, in his Alien Priories, presented to thirty benefices, and enjoyed an annual revenue of twenty thousand livres.—Among its English lands which were considerable, was the priory of Toft-Monks in our own immediate vicinity: the name, as you know, remains, though no traces of the building are now in existence.

The third abbey, that of St. Evrau or St. Evroul, called in Latin, Monasterium Uticense, was one of the most renowned throughout Normandy. The abbey dates its origin from St. Evroul himself, a nobleman, who lived in the reign of Childebert, and was attached to the palace of that monarch, "from which," to use the words of the chronicles, "he made his escape, as from shipwreck, and fled to the woods, and entered upon the monastic life."—The legend of St. Ebrulfus probably savors of romance, the almost inseparable companion of traditional, and particularly of monastic, history: it is safer, therefore, to be contented with referring the foundation of the monastery to the tenth century, when William Gerouis, after having been treacherously deprived of his sight and otherwise maimed, renounced the world; and, uniting with his nephews, Hugh and Robert de Grentemaisnil, brought considerable possessions to the endowment of this abbey. The abbey was at all times protected by the especial favor of the kings of France. No payment or service could be demanded from its monks; they acknowledged no master without their own walls, besides the sovereign himself; they were entitled to exemption from every kind of burthen; and they had the privilege of being empowered to castellate the convent, and to compel the people of the surrounding district to contribute their assistance for the purpose.

St. Evroul, however, principally claims our attention, as the sanctuary where Ordericus Vitalis, to use his own expressions, "delighted in obedience and poverty."—This most valuable writer was an Englishman; his native town being Attingesham, on the Severn, where he was born in the year 1075. He was sent to school at Shrewsbury, and there received the first rudiments, both of the humanities and of ecclesiastical education. In the tenth year of his age, his father, Odelerius, delivered the boy to the care of the monk Rainaldus. The weeping father parted from the weeping son, and they never saw each other more. Ordericus crossed the sea, and arrived in Normandy, an exile, as he describes himself, and "hearing, like Joseph in Egypt, a language which he understood not." In the eleventh year of his age, he received the tonsure from the hands of Mainerius, the abbot of St. Evroul. In the thirty-third year of his age, he was ordained a priest; and thenceforward his life wore away in study and tranquillity. Aged and infirm, he completed his Ecclesiastical History, in the sixty-seventh year of his age; and this great and valuable work ends with his auto-biography, which is written in an affecting strain of simplicity and piety.—The Ecclesiastical History of Ordericus is divided into parts: the first portion contains an epitome of the sacred and profane history of the world, beginning with the incarnation, and ending with Pope Innocent IInd. The second, and more important division, contains the history of Normandy, from the first invasion of the country, down to the year 1141.—Though professedly an ecclesiastical historian, yet Ordericus Vitalis is exceedingly copious in his details of secular events; and it is from these that his chronicle derives its importance and curiosity. It was first published by Duchesne, in his collection of Norman historians, a work which is now of rare occurrence, and it has never been reprinted.

Valuable materials for a new edition were, however, collected early in the eighteenth century, by William Bessin, a monk of St. Ouen; and these, before the revolution, were preserved in the library of that abbey. Bessin had been assisted in the task by Francis Charles Dujardin, prior of St. Evroul, who had collated the text, as published in the collection of Norman historians, with the original manuscript in his own monastery, to which latter Duchesne unfortunately had not access, but had been obliged to content himself with a copy, now in the Royal Library at Paris. It is to be hoped, that the joint labors of Bessin and Dujardin may still be in existence, and may come to light, when M. Liquet shall have completed the task of arranging the manuscripts in the public library at Rouen. The manuscript which belonged to St. Evroul, and was always supposed to be an autograph from the hands of Ordericus Vitalis himself, was discovered during the revolution among a heap of parchments, thrown aside as of no account, in some buildings belonging to the former district of Laigle. It is now deposited in the public library of the department of the Orne, but unfortunately, nearly half the leaves of the volume are lost. The earliest part of what remains is towards the close of the seventh book, and of this only a fragment, consisting of eight pages, is left. The termination of the seventh book, and the whole of the eighth are wanting. From the ninth to the thirteenth, both of these inclusive, the manuscript is perfect. A page or two, however, at the end of the work, which contained the author's life, has been torn out.—At the beginning of the sixteenth century, the manuscript was complete; for it is known that, at that time, a monk of St. Evroul made a transcript of it, which extended through four volumes in folio. These volumes were soon dispersed. Two of them found their way to Rouen, where they were kept in the library of St. Ouen: the other two were in that of the abbey of St. Maur de Glandefeuille, on the Loire. A third, though incomplete, copy of the original manuscript was also known to exist in France before the revolution. It formerly belonged to Coaslin de Camboret, Bishop of Metz, by whom it was presented, together with four thousand manuscripts, to the monks of St. Germain des Prés at Paris. But the greater part of the literary treasures of this abbey fell a prey to the flames in July, 1793, and it is feared that the copy of Ordericus perished at that time.

The original code from St. Evroul, was discovered by M. Louis Dubois, whom I have already mentioned in connection with the ruins of Neomagus. He is an antiquary of extensive knowledge and extraordinary zeal. His History of Lisieux, which he has long been preparing for the press, will be a work of great curiosity and interest. The publication of it is for the present suspended, whilst he superintends an edition of the Vaux-de-Vires, or Vaux de villes, of Olivier Basselin, an early Norman poet. Meanwhile, M. Dubois still continues his researches among the foundations of the ancient city, from which he has collected a number of valuable relics. Some of the most pleasant and instructive hours of my tour have been spent in his society; and, whilst it was under his guidance that I visited the antiquities of Lisieux, his learning assisted me in illustrating them. M. Dubois likewise possesses a large collection of original autograph letters, which I found much pleasure in perusing.

During the reign of Napoléon, he held the office of librarian of Alençon, a situation that afforded him the opportunity of meeting with many literary curiosities of this nature. Among others, which thus fell into his hands, was the following letter, written by the Princess Borghese, sister to the Emperor, and addressed to the Empress Marie-Louise, by whom it was received, while on a tour through the western departments. I annex a transcript of this epistle; for, although it has no immediate connection with the main subject of our correspondence, it yet is a very singular contribution towards the private history of the dynasty of Napoléon.—The odd mixture of caudle-cup compliment and courtly flattery, is sufficiently amusing. I have copied it, word for word, letter for letter, and point for point; for, as we have no other specimen of the epistles of her imperial highness, I think it right to preserve all the peculiarities of the original; and, by, way of a treat for the collectors of autographs, I have added a fac-simile of her signature.

> Madame et tres chere Sœur,
>
> je recois par le Prince Aldobrandini la lettre de V.M. et la belle tasse dont elle a daigné, le charger pour moi au nom de L'empereur, je remercie mille fois votre aimable bonté, et j'ose vous prier ma tres chere sœur d'être aupres de L'empereur l'interprete de ma reconnaissance pour cette marque de souvenir.—je fais parler beaucoup le Prince et la Princesse Aldobrandini sur votre santé, sur votre belle grossesse, je ne me lasse pas de les interroger, et je suis heureuse d'apprendre que vous vous portés tres bien, que rien ne vous fatigue, et que vous avés la plus belle grossesse qu'il soit possible de desirer, combien je desire chere sœur que tous nos vœux soient exaucés, ne croyés cependant pas que si vous nous donnés une petite Princesse je ne l'aimerais pas. non, elle nous serait chere, elle resemblerait a V.M. elle aurait sa douceur, son amabilité, et ce joli caractere qui la fait cherir de ceux qui out le bonheur de la Conaitre—mais ma chère sœur j'ai tort de m'apesantir sur les qualités dont serait douée cette auguste princesse, vous nous donnerés d'abord un prince un petit Roi de Rome, jugés combien je le desire nos bons toscans prient pour vous, ils vous aiment et je n'ai pas de peine a leur inspirer ce que je sens si vivement.

je vous remercie ma tres chere sœur de l'interest que vous prenez a mon fils, tout le monde dit qu'il ressemble a L'empereur. cela me Charme il est bien portant a present, et j'espere qu'il sera digne de servir sous les drapeaux de son auguste oncle.—adieu ma chere sœur soyés assés bonne pour Conserver un souvenir a une sœur qui vous est tendrement attachée. Napoléon ne cesse de lire la lettre pleine de bonté que V.M. a daigné lui ecrire, cela lui a fait sentir le plaisir qu'il y avait a savoir lire, et l'encourage dans ses etudes—je vous embrasse et suis,

Madame et tres chere Sœur

de V.M.

La plus attachée

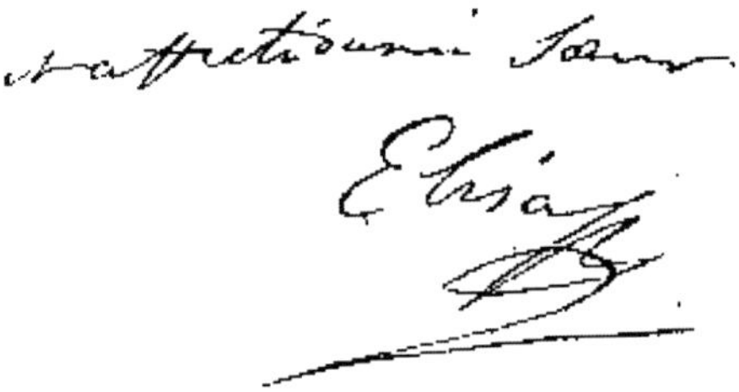

Pitti le 18 janvier 1811

Footnotes:

[66] See Magazin Encyclopédique, for 1802, III. p. 504.

[67] This transaction appears to have been peculiarly flagrant: a long detail of the circumstances, accompanied by several letters, very characteristic of the feeling and church-government of the times, is preserved in the Concilia Normannica, p. 520.—The account concludes in the following words:—"Exhorruit ad facinus, non Normannia solum et Anglia, quibus maledicta progenies notissima erat, sed et universa Gallia, et a singulis ad Apostolicum Paschalem delatum est. Nec tamen utrique simul ante quinquienniuin sordes de domo Dei propulsare prævaluerunt. Ceteris ferventiùs institit Yvo Carnotensis Antistes, conculcatæ disciplinæ ecclesiasticæ zelo succensus; in tantum ut Neustriacos Præsules quasi desides ac pusillanimes coarguere veritus non sit: sed ea erat Ecclesiæ sub ignavo Principe sors per omnia lamentabilis, ut ipsemet postmodum cum laude non invitus agnovit."

[68] Sandford, in his Genealogical History of the Kings of England, says, that this marriage was solemnized at Luxseul, in the county of Burgundy; but he refers for his authority to Ordericus Vitalis, by whom it is stated to have been at Luxovium, the name by which he always calls Lisieux; and he, in the same page, mentions the assembly of the nobles also held there.

[69] Annal, IV. p. 599.

LETTER XXIII
FRENCH POLICE—RIDE FROM LISIEUX TO CAEN—CIDER—GENERAL APPEARANCE AND TRADE OF CAEN—ENGLISH RESIDENT THERE

(*Caen, August*, 1818.)

Our reception at Caen has been somewhat inauspicious: we had scarcely made the few necessary arrangements at the hôtel, and seated ourselves quietly before the *caffé au lait*, when two gens-d'armes, in military costume, stalked without ceremony into the room, and, taking chairs at the table, began the conversation rather abruptly, with "Monsieur, vous êtes sous arrêt."—My companions were appalled by such a salutation, and apprehended some mistake; but the fact turned out to be, that our passport did not bear the signature of the mayor of Rouen, and that this ignorance of the regulations of the French police had subjected us to so unexpected a visit. It was too late in the day for the deficiency to be then supplied; and therefore, after a few expostulations, accompanied with observations, on their part, that we had the good fortune to have fixed ourselves at an *honnête hôtel*, and did not wear the appearance of suspicious persons, the soldiers took their leave, first exacting from me a promise, that I would present myself the next morning before the proper officer, and would in the meanwhile consider myself a prisoner upon my parole.

The impression which this occurrence could not fail to make upon our minds, was, that the object of the gens-d'armes had been either to extort from us money, or to shew their consequence; but I have since been led to believe that they did no more than their duty.—We have several acquaintance among the English who reside here, and we find from the whole of them, that the utmost strictness is practised in all matters relating to passports, and not less towards natives than foreigners. No Frenchman can quit his arrondissement unprovided with a passport; and the route he intends to take, and the distance he designs to travel, must also be specified. A week or two ago the prefect of the police himself was escorted back to Caen, between a couple of gens-d'armes, because he inadvertently paid a visit to a neighboring bathing-place without his passport in his pocket. This is a current story here: I cannot vouch for its authenticity; however it is certain, that since the discovery of the late plot contrived by the ultras, a plot whose existence is generally disbelieved, the French police is more than usually upon the alert.

When I presented myself at the Hôtel de Ville, to redeem my promise, a recent decree was pointed out to me, containing a variety of regulations which shew extraordinary uneasiness on the part of the government, and which would seem to indicate that they are in possession of intelligence respecting projects, that threaten the public tranquillity[70]. To judge from all official proceedings, it seems as if we were walking upon a smothered volcano, and yet we are told by every body that there is not the slightest room for apprehension of any kind.

This interruption has thrown me out of the regular course of my narration.—My last letter left me still at Lisieux, from which city to Caen the road lies through a tract of country altogether without interest, and in most places without beauty. During the first half of the ride, we could almost have fancied ourselves at home in Norfolk.—About this part of the way, the road descends through a hollow or dale, which bore the ominous name of "Coupe Gorge." When Napoléon was last in Normandy, he inquired into the origin of the appellation.—The diligences, he was answered, "had often been stopped and robbed in this solitary pass."—Napoléon then said, "If one person can be made to settle here, more will follow, for it is conveniently situated between two good towns. Let the prefect buy a little plot of ground and build a house upon it, and give it to an old soldier, upon condition that he shall constantly reside in it with his family." The orders of Napoléon were obeyed. The old soldier opened an inn, other houses arose round it, and the cut-throat pass is now thoroughly secure. The conductor and the post-boy tell the tale with glee whilst they drive through the hamlet; and its humble dwellings will perhaps recal the memory and fame of Napoléon Buonaparte when the brazen column of the grand army, and the marble arch of the Thuilleries, shall have been long levelled with the ground.—As to the character of the landscape, I must add, that though it makes a bad picture, there are great appearances of care in the agriculture, and of comfort in the population. The country, too, is sufficiently well wooded; and apple and pear trees every where take the place of the pollard oaks and elms of our hedge-rows.

Norman cider is famous throughout France: it is principally, however, the western part of the province that produces it. Throughout the whole of that district, the lower classes of the inhabitants scarcely use any other beverage. Vines, as I have already had occasion to mention, were certainly cultivated, in early times, farther to the north than they are at present. The same proofs exist of vineyards in the vicinity of Caen and Lisieux, as at

Jumieges. Indeed, towards the close of the last century, there was still a vineyard at Argence, only four miles south east of Caen; and a kind of white wine was made there, which was known by the name of Vin Huet. But the liquor was meagre; and I understand that the vineyard is destroyed.—Upon the subject of the early use of beer in Normandy, tradition is somewhat indistinct. The ancient name of one of the streets in Caen, rue de la Cervoisiere, distinctly proves the habit of beer-drinking; and, when Tacitus speaks of the beverage of the Germans, in his time, as "humor ex hordeo vel frumento in quandam similitudinem vini corruptus," it seems highly improbable but that the same liquor should have been in use among the cognate tribes of Gaul. Brito, however, expressly says of Flanders, that it is a place where,

"Raris sylva locis facit umbram, vinea nusquam:
Indigenis potus Thetidi miscetur avena,
Ut vice sit vini multo confecta labore."

And the same author likewise tells us, that the Normans of his time were cider-drinkers—

"... *Siceræque* potatrix
Algia tumentis ...
Non tot in autumni rubet Algia tempore *pomis*
Unde liquare solet *siceram* sibi *Neustria* gratam."

Huet is of opinion, that the use of cider was first introduced into Neustria by the Normans, who had learned it of the Biscayans, as these latter had done from the inhabitants of the northern coast of Africa.

We did not find the Norman cider at all palatable: it is extremely sour, hard, and austere. The inhabitants, however, say that this is not its natural character, but is attributable to the late unfavorable seasons, which have prevented the fruit from ripening properly.—The apple-tree and pear-tree in Normandy, far from being ugly, and distorted, and stunted in their growth, as is commonly seen in England, are trees of great beauty, and of extreme luxuriance, both in foliage and ramification. The *Coccus*, too, which has caused so much destruction among our orchards at home, is fortunately still unknown here.

The only place at which we stopped between Lisieux and Caen, was Croissanville, a poor village, but one that possesses a degree of historical interest, as the spot where the battle was fought between Aigrold, King of Denmark, and Louis d'Outremer, King of France; a battle which seated Richard Fearnought upon the throne of Normandy.—The country about Croissanville is an immense tract of meadow-land; and from it the Parisian market draws a considerable proportion of its supplies of beef. The cattle that graze in these pastures are of a large size, and red, and all horned; very unlike those about Caen, which latter are of small and delicate proportions, with heads approaching to those of deer, and commonly with black faces and legs.

From Croissanville to Caen the road passes through a dead flat, almost wholly consisting of uninclosed corn-fields, extending in all directions, with unvaried dull monotony, as far as the eye can reach. Buck-wheat is cultivated in a large proportion of them: the inhabitants prepare a kind of cake from this grain, of which they are very fond, and which is said to be wholesome. Tradition, founded principally upon the French name of this plant, sarrazin has given rise to a general belief, that buck-wheat was introduced int France by the Moors; but this opinion has, of late, been ably combate The plant is not to be found in Arabia, Spain, or Sicily; the countries mo particularly inhabited by Mahometans; and in Brittany, it still passes by t Celtic appellation, had-razin, signifying red-corn, of which words sarraz may fairly be regarded a corruption, as buck-wheat, in our own tongu ought unquestionably to be written beech-wheat; a term synonymous what it is called in Latin and German. The present name may well appe inexplicable, to those who are unacquainted with the Anglo-Saxon and cognate dialects.

In the midst of this level country, in which even apple-trees are scarc stands the ancient capital of Lower Normandy, extending from east t west in so long a line, that on our approach it appeared to cover as mucl ground as Rouen, which is in fact double its size.—From a distance, th view of Caen is grand; not only from the apparent magnitude of the towr but from the numerous spires and towers, that, rising from every part it, give it an air of great importance. Those of the abbeys of St. Stephe and the Trinity, at opposite extremities, constitute the principal features i the view.—The same favorable impressions continue when you enter th town. The streets are wide, and the houses of stone; and a stone city is pleasing sight to eyes long accustomed to the wooden buildings of Rouer Bernay, and Lisieux.—Besides, there is a certain degree of regularity i the construction of the buildings, and some care is taken in keeping then clean.—Lace-making is the principal occupation of females of the lower class in Caen and the neighborhood; the streets, as we passed along, were lined almost uninterruptedly on either side, with a row of lace-makers; and boys were not uncommonly working among the women. It is calculated tha not fewer than twenty thousand individuals, of all ages, from ten or twel years old and upwards, are thus employed; and the annual produce of th labor is estimated at one hundred and seventy thousand pounds sterli Caen lace is in high estimation for its beauty and quality, and is exported considerable quantities.

The present population of Caen amounts to about thirty-one thousand individuals. The town, no longer the capital of Lower Normandy, is still equally distinguished as the capital of the department of the Calvados. The prefect resides here; and the royal court of Caen comprises in its jurisdiction, not only the department more especially appertaining to it, but also those of the Manche and the Orne.—The situation of the town, though at the confluence of the Orne and the Odon, is not such as can be regarded favorable to extensive trade. The united rivers form a stream, which, though navigable at very high tides for vessels of two hundred tons burthen, will, on other occasions, admit only of much smaller ones; while the channel, nearer to its mouth, is obstructed by rocks that render the navigation difficult and dangerous. Many plans have been projected and attempted for the purpose of improving and enlarging the harbor, but little or no progress has yet been made. Vauban long since pointed out the mouth of the Orne as singularly well adapted for a naval station; and Napoléon, in pursuance of this idea, actually commenced the excavation of a basin under the walls of the town, and intended to deepen the bed of the river, thinking it best to make a beginning in this direction. All idea, however, of prosecuting such a plan is for the present abandoned.—Other engineers have proposed the junction of the Orne with the Loire by means of a canal, which would be of the greatest importance to France, not only by facilitating internal commerce, but by saving her vessels the necessity of coasting Capes Finisterre, and la Hogue, and thus enabling them to avoid a navigation, which is at all times dangerous, and in case of war peculiarly exposed.

For minor purposes, however, for mills and manufactories of different kinds, Caen is certainly well situated; being in almost every direction intersected with streams, owing to the repeated ramifications of the Odon, some of which are artificial, and of as early a date as the eleventh century. The same circumstance contributes materially to the pleasantness of the town; for the banks of the river are in many places formed into walks, and crowned by avenues of noble trees.

Female at Caen

The grand cours at Caen is almost as fine a promenade as that at Rouen. On Sunday evening it was completely crowded. The scene was full of life and gaiety, and very varied. All the females of the lower rank, and many of the higher orders, were dressed in the costume of the country, which commonly consists of a scarlet gown and deep-blue apron, or vice versâ. Their hair, which is usually powdered, is combed entirely back from every part of their faces, and tucked up behind. The snow-white cap which covers it is beautifully plaited, and has longer lappets than in the Pays de Caux. Mr. Cotman sketched the coiffure of the chamber-maid, at the Hôtel d'Espagne, in grand costume, and I send his drawing to you.—The men dress like the English; but do not therefore fancy that you or I should have any chance of being mistaken for natives, even if we did not betray ourselves by our accent. Here, as every where else, our countrymen are infallibly known: their careless slouching gait is sure to mark them; and the police keep a watchful eye upon them. Caen is at present frequented by the English: those indeed, who, like the Virgilian steeds, "stare loco nesciunt," seldom shew themselves in Lower Normandy; but above thirty British families have taken up their residence in this town: they have been induced to do so principally by the cheapness of living, and by the advantages held out for the education of their children. A friend of mine, who is of the number of temporary inhabitants, occupies the best house in the place, formerly the residence of the Duc d'Aumale; and for this, with the garden, and offices, and furniture of all kinds, except linen and plate, he pays only nine pounds a month. For a still larger house in the country, including an orchard and garden, containing three acres, well stocked with fruit-trees, he is asked sixty pounds from this time to Christmas. But, cheap as this appears, the expence of living at Coutances, or at Bayeux, or Valognes, is very much less.

Were I obliged to seek myself a residence beyond the limits of our own country, I never saw a place which I should prefer to Caen. I should not be tempted to look much farther before I said,

"Sis meæ sedes utinam senectæ:"—

The historical recollections that are called forth at almost every turn, would probably have some influence in determining my choice; the noble specimens of ancient architecture which happily remain, unscathed by wars and Calvinists and revolutions, might possibly have more; but the literary resources which the town affords, the pleasant society with which it abounds, and, above all, the amiable character of its inhabitants, would be my great attraction.—At present, indeed, we have not been here sufficiently long to say much upon the subject of society from our own experience; but the testimony we receive from all quarters is uniform in this point, and the civilities already shewn us, are of a nature to cause the most agreeable prepossessions. It is not our intention to be hurried at Caen; and I shall therefore reserve to my future letters any remarks upon its history and its antiquities. To a traveller who is desirous of information, the town is calculated to furnish abundant materials.

Footnotes:

[70] The following were among the articles of the decree:—"No individual to leave his arrondissement without a passport.—No person to receive a stranger in his house, or suffer one to quit it, without apprising the police.—The inhabitants to carry their arms of all kinds to the Hôtel de Ville.—No plays to be performed, except first approved by the officers of the police.—The manager of the theatre to give notice every Friday to the mayor, of the pieces intended to be acted the following week.—The actors to read nothing, and say nothing, which is not in the play.—The performance to begin precisely at six, and close at ten.—Only a certain interval to be allowed between the different pieces, or between the acts of each.—Every person to be uncovered, except the soldiers on duty.—No weapons of any kind, nor even sticks or umbrellas, to be taken into the theatre."

LETTER XXIV
HISTORIANS OF CAEN—TOWERS AND FORTIFICATIONS—CHÂTEAU DE LA GENDARMERIE—CASTLE—CHURCHES OF ST. STEPHEN, ST. NICHOLAS, ST. PETER, ST. JOHN, AND ST. MICHEL DE VAUCELLES

(*Caen, August,* 1818.)

France does not abound in topographical writers; but the history and antiquities of Caen have been illustrated with singular ability, by men to whom the town gave birth, and who have treated their subject with equal research and fidelity—these are Charles de Bourgueville, commonly called the Seigneur de Bras, and the learned Huet, Bishop of Avranches.

De Bourgueville was a magistrate of Caen, where he resided during almost the whole of the sixteenth century. The religious wars were then raging; and he relates, in a most entertaining and artless manner, the history of the events of which he was an eye-witness. His work, as is justly observed by Huet, is a treasure, that has preserved the recollection of a great variety of the most curious details, which would otherwise have been neglected and forgotten. Every page of it is stamped with the character of the author—frankness, simplicity, and uprightness. It abounds in sound morality, sage maxims, and proofs of excellent principles in religion and politics; and, if the writer occasionally carries his naïveté to excess, it is to be recollected that the book was published when he was in his eighty-fifth year, a period of life when indulgence may reasonably be claimed. He died four years subsequently, in 1593.—In Huet's work, the materials are selected with more skill, and are digested with more talent. The author brought to his task a mind well stored with the learning requisite for the purpose, and employed it with judgment. But he has confined himself, almost wholly, to the description of the town; and the consequence is, that while the bishop's is the work most commonly referred to, the magistrate's is that which is most generally read. The dedication of the former to the town of Caen, does honor to the feelings of the writer: the portrait of the latter, prefixed to his volume, and encircled with his quaint motto, "L'heur de grace use l'oubli," itself an anagram upon his name, bespeaks and insures the good will of the reader.

The origin of Caen is uncertain.—Its foundation has been alternately ascribed to Phoenicians, Romans, Gauls, Saxons, and Normans. The earliest historical fact connected with the town, is recorded in an old chronicle of Normandy[71], written in 1487, by William de Talleur, of Rouen. The author, in speaking of the meeting between Louis d'Outremer, King of France, and Richard Ist, Duke of Normandy, about the year 945, enumerates Caen among the good towns of the province. Upon this, Huet observes that, supposing Caen to have been at that time only recently founded, it must have acquired importance with much rapidity; for, in the charter, by which Richard IIIrd, Duke of Normandy, granted a dowery to Adela, daughter of Robert, King of France, whom he married in 1026, Caen is not only stated as one of the portions of the dower, but its churches, its market, its custom-house, its quay, and its various appurtenances are expressly mentioned; and two hundred years afterwards, Brito in his Philippiad, puts Caen in competition with Paris,

"Villa potens, opulenta situ, spatiosa, decora,
Fluminibus, pratis, et agrorum fertilitate,
Merciferasque rates portu capiente marino,
Seque tot ecclesiis, domibus et civibus ornans,
Ut se Parisio vix annuat esse minorem."—

Caen is designated in Duke Richard's charter, by the appellation of "in Bajocensi comitatu villa quæ dicitur *Cathim*, super fluvium Olnæ."—From *Cathim*, came *Cahem*; and *Cahem*, in process of time, was gradually softened into *Caen*. The elision that took place in the first instance, is of a similar nature to that by which the Italian words *padre* and *madre*, have been converted into *père* and *mère*; and the alteration in the latter case continued to be indicated by the diæresis, which, till lately, separated the two adjoining vowels.—Towards the latter part of the eleventh century, Caen is frequently mentioned by the monkish historians, in whose Latin, the town is styled *Cadomus* or *Cadomum*.—And here ingenious etymologists have found a wide field for conjecture: Cadomus, says one, was undoubtedly founded by Cadmus; another, who hesitates at a Phoenician antiquity, grasps with greater eagerness at a Roman etymon, and maintains that *Cadomus* is a corruption from *Caii domus*, fully and sufficiently proving that the town was built by Julius Cæsar.

Robert Wace states, in his *Roman de Rou*, that, at the time immediately previous to the conquest of England, Caen was an open town.—

"Encore ert Caen sans Châtel,

N 'y avoit mur, ny quesnel."—

And Wace is a competent witness; for he lived during the reign of Henry Ist, to whom he dedicated his poem. Philip de Valois, in 1346, allowed the citizens to surround the town with ditches, walls, and gates. This permission was granted by the king, on the application of the inhabitants, Caen, as they then complained, being still open and unfortified. Hence, the fortifications have been considered to be the work of the fourteenth century, and, generally speaking, they were unquestionably, of that time; but it is equally certain, that a portion was erected long before.

A proof of the antiquity of the fortifications may perhaps be found in the name of the tower called la Tour Guillaume le Roi, which stands immediately behind St. Peter's, and was intended to protect the river at the extremity of the walls, dividing the town from the suburb of Vaugeux. This tower is generally supposed to be the oldest in the fortifications. Its masonry is similar to that of the wall with which it is connected, and which is known to have been built about the same time as the abbey of St. Stephen. The appearance of it is plain, massy, and rugged; and it forms a picturesque object. Such also is the Tour au Massacre, which is situated at the confluence of the Orne and Odon. The tower in question is said to have received its gloomy title from a massacre, of which our countrymen were guilty, at the time when the town was taken in 1346. There is, however, reason to believe that this tale is a mere fiction. Huet, at the same time that he does not venture so far to oppose popular belief, as altogether to deny the truth of the story of the massacre, adds, that the original name of the tower was la Tour Machart, and suspects its present appellation to be no more than a corruption of the former one. Renauld Machart was bailiff of Caen two years prior to the capture of the place by Edward IIIrd; and the probability is, that the tower was erected by him in those times of alarm, and thus took his name. It has been supposed that the figure sculptured upon it, may also be intended for a representation of Machart himself.

Caen contains another castellated building, which might easily mislead the studious antiquarian. The Château de Calix, as it is sometimes called, is situated at the extremity of the suburb known by that name; and the curious inhabitants of Caen usually suppose that it was erected for the purpose of commanding the river, whilst it flowed in its ancient, but now deserted, bed; or, at least, that it replaces such a fortification. According to the learned Abbé de la Rue, however, and he is a most competent authority, no real fortification ever existed here; but the castle was raised in conformity to the caprice of Girard de Nollent, the wealthy owner of the property, who flourished towards the beginning of the sixteenth century.—Girard de

Nollent's mansion is now occupied by a farmer. It has four fronts. The windows are square-headed, and surrounded by elegant mouldings; but the mullions have been destroyed. One medallion yet remains over the entrance; and it is probable that the walls were originally covered with ornaments of this kind. Such, at least, is the case with the towers and walls, which, surrounding the dwelling, have given it a castellated aspect. The circular tower nearest the gate forms the subject of the accompanying sketch: it is dotted on all sides with busts in basso-relievo, enclosed in medallions, and of great diversity of character. One is a frowning warrior, arrayed in the helmet of an emperor of the lower empire; another, is a damsel attired in a ruff; a third, is a turbaned turk. The borders of the medallions are equally diversified: the cordelière, well known in French heraldry, the vine-leaf, the oak-leaf, all appear as ornaments. The battlements are surmounted with two statues, apparently Neptune, or a sea-god, and Hercules. These heathen deities not being very familiar to the good people of Caen, they have converted them, in imagination, into two gens-d'armes, mounting guard on the castle; and hence it is frequently called the Château de la Gendarmerie. Some of the busts are accompanied by inscriptions—"Vincit pudicitiam mors;" "Vincit amor pudicitiam;" "Amor vincit mortem;" and all seem to be either historical or allegorical. The battlements of the curtain-wall are ornamented in the same manner. The farther tower has less decoration, and is verging to decay. I have given these details, because the castle of Calix is a specimen of a style of which we have no fair parallel in England, and the workmanship is far from being contemptible.

In the Rue St. Jean is a house with decorations, in the same style, but more sumptuous, or, perhaps I ought rather to say, more perfect. Both of them are most probably of nearly the same date: for it was principally during the reigns of Charles VIIIth and Louis XIIth, that the practice prevailed in France, of ornamenting the fronts of houses with medallions. The custom died away under Francis Ist.

I must now return to more genuine fortifications.—When the walls of Caen were perfect, they afforded an agreeable and convenient promenade completely round the town, their width being so great, that three persons might with ease walk abreast upon them. De Bourgueville tells us that, in his time, they were as much frequented as the streets; and he expatiates with great pleasure upon the gay and busy prospect which they commanded,

The castle at Caen, degraded as it is in its character by modern innovation, is more deserving of notice as an historical, than as an architectural, relic. It still claims to be ranked as a place of defence, though it retains but few of its original features. The spacious, lofty, circular towers, known by the names of the black, the white, the red, and the grey horse, which flanked its ramparts, have been brought down to the level of the platform. The dungeon tower is destroyed. All the grandeur of the Norman castle is lost; though the width of its ditches, and the thickness of its walls, still testify its ancient strength. I doubt whether any castle in France covers an equal extent of ground. Monstrelet and other writers have observed, that this single fortress exceeded in size the towns of Corbeil or of Montferrand; and, indeed, there are reasons for supposing that Caen, when first founded, only occupied the site of the present castle; and that, when it became advisable to convert the old town into a fortress, the inhabitants migrated into the valley below. Six thousand infantry could be drawn up in battle-array within the outer ballium; and so great was the number of houses and of inhabitants enclosed within its area, that it was thought expedient to build in it a parochial church, dedicated to St. George, besides two chapels.

One of the chapels is still in existence, though now converted to a store-house; and the Abbé de la Rue considers it as an erection anterior to the conquest, and, belonging to the old town of Caen. Its choir is turned towards the west, and its front to the east.—The religious edifices upon the continent do not preserve the same uniformity as our English ones, in having their altars placed in the direction of the rising sun; but this at Caen is a very remarkable instance of the position of the entrance and the altar being completely reversed[72]. The door-way is a fine semi-circular arch: the side pillars supporting it are very small, but the decorations of the archivolt are rich: they consist principally of three rows of the chevron moulding, enclosed within a narrow fillet of smaller ornaments, approaching in shape to quatrefoils. Collectively, they form a wide band, which springs from flat piers level with the wall, and does not immediately unite with the head of the inner arch. The intermediate space is covered by a reticulated pattern indented in the stone. Above the entrance is a window of the same form, its top encircled by a broad chequered band, a very unusual accompaniment to this style of architecture. The front of the chapel presents in other respects, a flat uniform surface, unvaried, except by four Norman buttresses, and a string-course of the simplest form, running round the whole building, at somewhat less than mid-height. The sides of the chapel are lighted by a row of circular-headed windows, with columns in the angles; and between these windows are buttresses, as in the chapel of the lazar-house of St. Julien, at Rouen.

Huet endeavours to prove that the first fortress which was built at Caen, was erected by William the Conqueror, who frequently resided here with his Queen Matilda, and who was likely to find some protection of this nature desirable, as well to guard his royal residence against the mutinous disposition of the lords of the Bessin, as to command the navigation of the Orne. The castle was enlarged and strengthened by his son Henry; but it is believed that the four towers, just mentioned, and the walls surrounding the keep, were added by our countrymen, during that short period when the Norman sceptre was again wielded by the descendants of the Norman dukes. Under Louis XIIth and Francis Ist, the whole of the castle, but particularly the dungeon, underwent great repairs, by which the original form of the structure was entirely changed.—From that period history is silent respecting the fortress. I cannot, however, take leave of it without reminding you, that Sir John Fastolf, whilom our neighbour at Castor, was for some time placed in command here, as Lieutenant to the Regent Duke of Bedford. You, who are acquainted with the true character of the knight, need scarcely be told, that even his enemies concur in bearing testimony to his ability, his vigilance, and his valor: it is to be regretted that he has not met with equal justice at home. Not one individual troubles himself about history, whilst a thousand read the drama; and the stains which Shakspeare's pen has affixed to the name of Fastolf, are of a nature never to be wiped away; thus disproving the distich of the satyrist, who indeed, by his own works, has effectually falsified his own maxim, that—

> "Truth will survive when merry jokes are past;
> For rising merit must buoy up at last."

As usual, the buildings dedicated to religion are far more numerous and valuable than the relics of military architecture. Of these, the first which salutes the stranger who enters by the great high road, is the Hôtel Dieu, which is almost intact and unaltered. The basement story contains large and deep pointed arches, ornamented with the chevron moulding, disposed in a very peculiar manner.—From the style of the building, there is every reason to believe that it is of the beginning of the thirteenth century, at which time William, Count of Magneville, appropriated to charitable purposes the ground now occupied by this hospital, and caused his donation to be confirmed by a bull from Pope Innocent IIIrd, dated in April, 1210.

The abbeys, the glories of Caen, will require more leisure: at present let us pass on to the parochial churches. Of these, the most ancient foundation is *St. Etienne le Vieil*; and tradition relates that this church was dedicated by St. Renobert, bishop of Bayeux, in the year 350.—But, though the present edifice may stand upon the site of an ancient one, there would be little risk in affirming, that not one stone of it was laid upon another till after the year 1400. The building is spacious, and its tower is not devoid of beauty. The architecture is a medley of debased gothic and corrupted Roman; but the large pointed windows, decorated by fanciful mouldings and scroll-work, have an air of richness, though the component parts are so inharmonious.

Attached to the wall of the choir of this church is still to be seen an equestrian statue[73], part of the celebrated group supposed to represent William the Conqueror making his triumphal entry into Caen. A headless horse, mounted by a headless rider, and a figure, which has lost all shape and form, beneath the feet of the steed, are all that now remain; but De Bourgueville, who knew the group when perfect, says, that there likewise belonged to it a man and woman upon their knees, as if seeking some explanation for the death of their child, or rather, perhaps, in the act of imploring mercy.—I have already pointed out the resemblance between these statues and the bas-relief, of which I have sent you a sketch from St. Georges. One of the most learned antiquaries of the present time has found a prototype for the supposed figure of the Duke, among the sculptures of the Trajan column. But this, with all due deference, is far from a decisive proof that the statue in question was not intended for William. Similar adaptations of the antique model, "mutato nomine," frequently occur among the works of the artists of the middle ages; and there is at least a possibility that, had the face been left us, we might have traced some attempt at a portrait of the Norman Duke. Upon the date of the sculpture, or the style of the workmanship, I dare not venture an opinion. There are antiquaries, I know, (and men well qualified to judge,) who believe it Roman: I have heard it pronounced from high authority, that it is of the eleventh century, others suspect that it is Italian, of the thirteenth or fourteenth centuries; whilst M. Le Prevost and M. De Gerville maintain most strenuously that it is not anterior to the fifteenth. De Bourgueville certainly calls it "une antiquité de grand remarque;" but we all know that any object which is above an hundred years old, becomes a piece of antiquity in the eye of an uncritical observer; and such was the good magistrate.

The church of St. Nicholas, now used as a stable, was built by William the Conqueror, in the year 1060, or thereabouts. Desecrated as it is, it remains entire; and its interior is remarkable for the uniformity of the plan, the symmetry of the proportions. All the capitals of the pillars attached to the walls are alike; and those of the arches, which very nearly resemble the others, are also all of one pattern. In the side-aisles there is no groining, but only cross vaulting. The vaulting of the nave is pointed, and of late introduction. Round the choir and transepts runs a row of small arches, as in the triforium.—The west end was formerly flanked by two towers, the southern of which only remains. This is square, and well proportioned: each side contains two lancet windows. The lower part is quite plain, excepting two Norman buttresses. The whole of the width of the central compartment, which is more than quadruple that of either of the others, is occupied below by three circular portals, now blocked up.—Above them are five windows, disposed in three tiers. In the lowest are two not wider than loop-holes: over these two others, larger; another small one is at the top. All these windows are of the simplest construction, without side pillars or mouldings.—The choir of the church ends in a semi-circular apsis, divided into compartments by a row of pillars, rising as high as the cornice: in the intercolumniation are windows, and under the windows small arches, each of which has its head hewn out of a single stone.—The roof of the choir is of stone, and the pitch of it is very high.

Here, then, we have the exact counterpart of the Irish stone-roofed chapels, the most celebrated of which, that of Cormac, in Cashel Cathedral, appears, from all the drawings and descriptions I have seen of it, to be altogether a Norman building. Ledwich asserts that "this chapel is truly Saxon, and was erected prior to the introduction of the Norman, and gothic styles[74]." If, we agree with him, we only obtain a proof that there is no essential difference between Norman and Saxon architecture; and this proposition, I believe, will soon be universally admitted. We now know what is really Norman; and a little attention to the buildings in the north of Germany, may terminate the long-debated questions, relative to Saxon architecture and the origin of the stone-roofed chapels in the sister isle.

In the burial-ground that surrounds the church of St. Nicholas, are several monumental inscriptions, all of them posterior to the commencement of the reign of Napoléon, and all, with one single exception, commemorative of females. The epitaphs are much in the same tone as would be found in an English church-yard. The greater part, however, of the tomb-stones, are uninscribed. They are stone coffins above-ground, sculptured with plain crosses, or, where they have been raised to ecclesiastics, with an addition of some portion of the sacerdotal dress.

Among the churches of comparatively modern erection, St. Peter deserves most attention. From every part of the town and neighborhood, its lofty spire, towering above the surrounding buildings, forces itself upon your view. It is not easy to carry accurate ideas of height in the memory; but, as far as recollection will serve me, I should say that its elevation is hardly inferior to that of the spire of Salisbury cathedral. I have no hesitation in adding, that the proportions of the tower and spire of the church at Caen, are more pleasing. Elegance, lightness, and symmetry, are the general characters of the whole, though the spire has peculiar characters of its own.—The tower, though built a century later than that of Salisbury, is so much less ornamented, that it might be mistaken for an earlier example of the pointed style. The lowest story is occupied wholly by a portal: the second division is surrounded by pointed arches, beneath crocketed gables: the third is filled by four lancet arches, supported by reeded pillars, so lofty, that they occupy nearly two-thirds of the entire height of the tower. The flanking arches are blanks: the two middle ones are pierced into windows, divided by a central mullion. The balustrade at the top of the tower is of a varied pattern, each side exhibiting a different tracery. Eight crocketed pinnacles are added to the spire, which is octangular, and has a row of crockets at each angle. From the base to the summit it is encircled, at regular distances, with broad bands of stone-work, disposed like scales; and, alternating with the bands, are perforations in the form of cinquefoils, quatrefoils, and trefoils, diminishing as the spire rises, but so disposed, that the light is seen distinctly through them. The effect of these perforations was novel and very pleasing.

This tower and spire were built in the year 1308, under the directions of Nicolle L'Anglois, a burgher of Caen, and treasurer of the church.—How far we are at liberty to infer from his name, as Ducarel does, that he was an Englishman, may admit of some doubt. He was buried here; and De Bourgueville has preserved his epitaph, which recounts among his other merits, that

> "Et par luy, et par sa devise
> Fut la tour en sa voye mise
> D 'estre faicte si noblement."—

But the name of the architect who was employed is unrecorded.—The rest of the church was erected at different periods: the northern aisle in 1410; the opposite one some time afterwards; and the eastern extremity, with the vaulted roof of the choir and aisles, in 1021.—With this knowledge, it is not difficult to account for the diversity of styles that prevails in the building.—The western front contains much good tracery, and well disposed, apparently as old as the tower.—The exterior of the east end, with its side-chapels, is rather Italian than gothic.—The interior is of a purer style: the five arches forming the apsis are perhaps amongst the finest specimens of the luxuriant French gothic: roses are introduced with great effect amongst the tracery and friezes, with which the walls are covered. The decorations of the chapels round the choir, although they display a tendency towards Italian architecture, are of the most elaborate arabesque. The niches are formed by escalop shells, swelling cylinders of foliage, and scrolls: some of the pendants from the roofs are of wonderfully varied and beautiful workmanship.—The nave has nothing remarkable, saving the capital of one of the side pillars. Its sculptures, with the exception of one mutilated group, have been drawn by Mr. Cotman.—The subjects are strangely inappropriate, as the ornaments of a sacred edifice. All are borrowed from romance.—Aristotle bridled and saddled by the mistress of Alexander. Virgilius, or, as some say, Hippocrates, hanging in the basket. Lancelot crossing the raging flood.—The fourth, which is not shewn in the sketch, is much defaced, but seems to have been taken from the Chevalier et la Charette. According to the usual fate of ancient sculpture, the marguilliers of the parish have so sadly encumbered it with white-wash, that it is not easy to make out the details; and a friend of mine was not quite certain whether the bearded figure riding on the lion, was not a youthful Cupid. No other of the capitals has at present any basso-relievo of this kind; but I suspect they have been chopped off. The church suffered much from the Calvinists; and afterwards, during the revolution, when most of the bas-reliefs of the portal were destroyed.

Church of St. Jean at Caen

The neighboring church of St. John appears likewise to be the work of the fourteenth and fifteenth centuries. This building and St. Peter's agree in general character: their towers are nearly the counterparts of each other. But, in St. John's, the great tower is placed at the west end of the edifice, the principal portal being beneath it. This is not very usual in the Norman-gothic churches, though common in England. The tower wants a spire; and, at present, it leans considerably out of the perpendicular line, so that some apprehensions are entertained for its safety. It was originally intended that the church should also be surmounted by a central tower; and, as De Bourgueville says, the beginning was made in his time; but it remains to the present day incomplete, and has not been raised sufficiently high to enable us to form a clear idea of the design of the architect, though enough remains to shew that it would have been built in the Romanizing-gothic style.—The inside is comparatively plain, excepting only the arches in the lower open part of the tower. These are richly ornamented; and a highly-wrought balustrade runs round the triforium, uniform in its pattern in the nave and choir, but varying in the transepts.—In the other ecclesiastical buildings at Caen, we saw nothing to interest us.—The chapel of St. Thomas l'Abattu, which, according to Huet, "had existed from time immemorial," and which, to judge from Ducarel's description and figure, must have been curious, has now entirely disappeared.

In the suburb of Vaucelles, the church of St. Michael contains some architectural features of great curiosity[75]. The circular-headed arches in the short square tower, and in a small round turret that is attached to it, are unquestionably early Norman, and are remarkable for their proportions, being as long and as narrow as the lancet windows of the following æra. It would not be equally safe to pronounce upon the date of the stone-roofed pyramid which covers this tower. The north porch is entered by a pointed arch, which, though much less ornamented, approaches in style to the southern porch of St. Ouen, and, like that, has its inner archivolt fringed with pendant trefoils. The wall above the arch rises into a triangular gable, entirely covered with waving tracery, the only instance of the kind which I have seen at Caen.

Footnotes:

[71] Huet, Origines de Caen, p. 12.

[72] Upon this subject, Huet has an extraordinary observation, (Origines de Caen, p. 186.) "that, in the early times of Christianity, it was customary for all churches to front the east or north, or some intermediate point of the

compass."—So learned and careful a writer would scarcely have made such a remark without some plausible grounds; but I am at a loss where to find them. Bingham, in his Origines Eccleslasticæ, I. p. 288, says, "that churches were so placed, that the front, or chief entrances, were towards the west, and the sanctuary or altar placed towards the east;" and though he adduces instances of a different position, as in the church of Antioch, which faced the east, and that of St. Patrick, at Sabul, near Down in Ulster, which stood from north to south, he cites them only as deviations from an established practice.

[73] Cotman's Architectural Antiquities of Normandy, t. 20.

[74] Antiquities of Ireland, p. 151.

[75] See Cotman's Architectural Antiquities of Normandy, t. 18, 19.

LETTER XXV
ROYAL ABBEYS OF THE HOLY TRINITY AND ST. STEPHEN—FUNERAL OF THE CONQUEROR, EXHUMATION OF HIS REMAINS, AND DESTRUCTION OF HIS MONUMENT

(*Caen, August*, 1818.)

The two royal abbeys of Caen have fortunately escaped the storms of the revolution. These buildings are still standing, an ornament to the town, and an honor to the sovereign who caused them to be erected, as well as to the artist who planned, and to the age which produced them. As models of architecture they are the same land-marks to the history of the art in Lower Normandy, as the church of St. Georges is in the upper division of the province. Their dates are equally authenticated; and the characteristic features in each are equally perfect.

Both these noble edifices rose at the same time, and from the same motive. William the Conqueror, by his marriage with Matilda, daughter of Baldwin, Earl of Flanders, had contracted an alliance proscribed by the degrees of consanguinity. The clergy inveighed against the union; and they were supported in their complaints by Lanfranc, then resident at Bec, whose remonstrances were so uncourtly and strenuous, that the duke banished him from the province. It chanced that the churchman, while in the act of obedience to this command, met the sovereign. Their interview began with recriminations: it ended with reconciliation; and Lanfranc finally engaged to undertake a mission to the supreme Pontiff, who, considering the turbulent disposition of the Normans, and that a better end was likely to be answered by peaceable than by hostile measures, consented to grant the necessary dispensation. At the same time, by way of penance, he issued an injunction that the royal pair should erect two monasteries, the one for monks, the other for nuns. And in obedience to this command, William founded the abbey of St. Stephen, and Matilda, the abbey of the Holy Trinity; or, as they are usually called at Caen, l'abbaye aux hommes, and l'abbaye aux dames.

The approach to the monastery of the Trinity is through a spacious gate-tower, part of the original structure. Over the rent and shapeless door-way are three semi-circular arches, upon the capitals of which is distinctly observable the cable-moulding, and along the top of the tower runs a line of the same toothed ornament, remarked by Ducarel at Bourg-Achard, and stated by him to have been considered peculiar to Saxon architecture[76]. The park that formerly environed the abbey retains its character, though abandoned to utter neglect. It is of great extent, and is well wooded. The monastic buildings, which are, as usual, modern, are mostly perfect.—A ruined wall nearly in front of the church, with a chimney-piece, perhaps of Norman workmanship, belonged to the old structure. Such part of the chimney wall as was exposed to the flame is built of large tiles, placed diagonally. All other vestiges of the ancient apartments have been removed.

The noble church[77] is now used as a work-house for the department. At the revolution it became national property, and it remained unappropriated, till, upon the institution of the Legion of Honor, Napoléon applied it to some purpose connected with that body, by whom it was lately ceded for it present object. But, if common report may be credited, it is likely soon to revert to its original destination. The restoration may be easily effected, as the building has sustained but little injury. A floor has been thrown across the nave and transept, dividing them into two stories; but in other respects they are unaltered, and divine service is still performed in the choir.

A finer specimen of the solid grandeur of Norman architecture is scarcely to be found any where than in the west front of this church. The corresponding part of the rival abbey of St. Stephen is poor when compared to it; and Jumieges and St. Georges equally fail in the comparison. In all of these, there is some architectural anomaly: in the Trinity none, excepting, indeed, the balustrade at the top of the towers; and this is so obviously an addition of modern times, that no one can be misled by it. This balustrade was erected towards the beginning of the seventeenth century, when the oval apertures and scrolls seen in Ducarel's print were introduced. Anciently the towers were ornamented with very lofty spires. According to some accounts, these were demolished, because they served as land-marks to the English cruizers, being seen far out at sea; but other accounts state, that the spires were pulled down by Charles, King of Navarre, who was at war with his namesake, Charles Vth, then Dauphin and Regent. The abbey at that time bore the two-fold character of nunnery and fortress.—Strangely inconsistent as this union may appear, the fact is undoubted. Even now a portion of the fosses remains; and the gate-way indicates an approach to a fortified place. Ancient charters likewise expressly recognize the building in both capacities: they endow the abbey for the service of God; and they enjoin the inhabitants of the adjacent parishes to keep the fortifications in repair against any assaults of men. Nay, letters patent, granted by Charles

Vth, which fix the salary of the captain of the Fort of the Trinity, at Caen, at one hundred francs per annum, are yet extant.

I shall attempt no description of the west front of this monastery, few continental buildings being better known in England. The whole remains as it was in the time of Ducarel, except that the arches of entrance are blocked up, and modern windows have been inserted in the door-ways.—The north side of the church is quite concealed by the cloisters and conventual buildings. The southern aisle has been plastered and patched, and converted into a range of work-shops, so that its original elevation is wholly obliterated. But the nave, which rises above, is untouched by innovation. The clerestory range is filled by a row of semi-circular headed windows, separated by intervening flat buttresses, which reach to the cornice. Each buttress is edged with two slender cylindrical pilasters; and each window flanked by two smaller arches, whose surfaces are covered with chequer-work. The arch of every window has a key-stone, formed by a grotesque head.—Above the whole is a corbel-table that displays monsters of all kinds, in the form of beasts, and men scarcely less monstrous.—The semi-circular east end is divided in its elevation into three compartments. The lower contains a row of small blank arches: in each of the other two is a window, of a size unusually large for a Norman building, but still without mullions or tracery; its sides ornamented with columns, and its top encircled with a broad band of various mouldings. The windows are separated by cylindrical pillars, instead of buttresses.—In the upper part of the low central tower are some pointed arches, the only deviations of style that are to be found in the building. To the extremity of the southern transept has been attached a Grecian portico, which masks the ancient portal. Above is a row of round arches, some of which are pierced into windows.

Of the effect of the nave and transept within, it is difficult now to obtain a correct idea, the floor intervening to obstruct a general view.—High arches, encircled with the embattled moulding below; above these, a wide billeted string-course, forming a basis for a row of smaller arches, without side-pillars or decoration of any kind; then another string-course of different and richer patterns; and over this, the triforium, consisting also of a row of small arches, supported by thick pillars;—such is the elevation of the sides of the nave; and the same system is continued with but small variation in the transepts. But, notwithstanding the general uniformity of the whole, no two compartments are precisely alike; and the capitals are infinitely varied. It is singular to see such a playfulness of ornament in a building, whose architect appears, at first view, to have contemplated only grandeur and solidity.—The four arches which support the central tower are on a magnificent scale. The archivolts are encircled by two rows of lozenged squares, indented in the stone. The rams, or rams' heads, upon the capitals of these piers, are peculiar. The eastern arch rises higher than the rest, and is obtusely pointed; yet it seems to be of the same date with its circular companions.—So exquisite, however, is the quality of the Caen stone, that no opinion drawn from the appearance of the material, ought to be hazarded with confidence. Seven centuries have elapsed since this church was erected, and there is yet no difference to be discovered in the color of the stone, or the sharpness of the work; the whole is as clean and sharp as if it were but yesterday fresh from the chisel. The interior of the choir has not been divided by the flooring; and the eastern extremity, which remains perfect, shews the original design. It consists of large arches, disposed in a double tier, so as to correspond with the windows of the apsis, and placed at a short distance from the wall; but without any Lady-Chapel beyond. The pillars that support these arches are well proportioned: the sculptures on their capitals are scarcely less grotesque than those at St. Georges; but, barbarous as they are, the corners of almost every capital are finished with imitations, more or less obvious, of the classical Ionic volute.—Among the sculptures is a head resting upon two lions, which has been fancied to be a representation of the Conqueror himself; whilst a faded painting of a female, attired as a nun, on the north side of the altar, is also commonly entitled a portrait of the foundress.—Were any plausible reason alleged for regarding the picture as intended to bear even an imaginary resemblance to Matilda, I would have sent you a copy of it; but there appear no grounds to consider it as authentic.—Willing, however, to contribute a mark of respect to a female, styled by William of Malmesbury, "fæminam prudentiæ speculum, pudoris culmen," and, by way of a companion to the rough sketch of her illustrious consort, in the initial letter in the library at Rouen, I add the fac-simile of a seal, which, by the kindness of a friend has fallen into my hands. It has been engraved before, but only for private distribution; and, if a suspicion should cross your mind, that it may have belonged to the Empress Maud, or to Matilda, wife to Stephen, I can only bespeak your thanks to me, for furnishing you with a likeness of any one of these ladies.

Matilda was interred in the middle of this choir; and, according to Ordericus Vitalis, a monument of exquisite workmanship, richly ornamented with gold and precious stones, and bearing a long inscription in letters of gold, was raised to her memory. Her effigy was afterwards added to the monument; the whole of which was destroyed in 1652, by the Calvinists, who tore open the Queen's coffin, and dispersed her remains. After a lapse of an hundred and forty years, the royal bones were again collected, and deposited in this church. At the same time, the splendid monument was replaced by a plain altar-tomb, which existed till the revolution, when all was once more swept away. The marble slab, inscribed with the original epitaph, alone remained entire, and was carried to the abbey church of St. Stephen's, where it still forms a part of the pavement in a chapel. The letters are finely sculptured and perfectly sharp. However, it is not likely to continue there long; for Count de Montlivault, the prefect of the department, has already caused a search to be made for Matilda's remains, and he intends to erect a third monument to her memory. The excavations for this purpose have hitherto been unsuccessful: the Count met with many monumental stones, and many coffins of various kinds, but none that could be mistaken for the desired object; for one of the inscriptions on the late monument expressly states, that the Queen's bones had been wrapped in a linen cloth, and enclosed in a leaden box.

The inquiry, however, will not be discontinued[78]: there are still hopes of success, especially in the crypt, which corresponds in its architecture with the church above. It is filled with columns placed in four ranges, each standing only four feet from the other, all of elegant proportions, with diversified capitals, as those in the choir.—Round it runs a stone bench, as in the subterraneous chapel in St. Gervais, at Rouen.

Founded by a queen, the abbey of the Trinity preserved at all times a constitution thoroughly aristocratical. No individual, except of noble birth, was allowed to take the veil here, or could be received into the community. You will see in the series of the abbesses the names of Bourbon, Valois, Albret, Montmorenci, and others of the most illustrious families in France. Cecily, the Conqueror's eldest daughter, stands at the head of the list. According to the *Gallia Christiana,* she was devoted by her parents to this holy office, upon the very day of the dedication of the convent, in July 1066.

The black marble slab which covered her remains, was lately discovered in the chapter-house. A crozier is sculptured upon it. It is delineated in a very curious volume now in the possession of the Abbé de la Rue, which contains drawings of all the tombs and inscriptions that formerly existed in the abbey.

The annual income of the monastery of the Trinity is stated by Gough, in his Alien Priories, at thirty thousand livres, and that of the monastery of St. Stephen, at sixty thousand; but Ducarel estimates the revenue of the former at seventy thousand, and of the latter at two hundred thousand; and I should not doubt but that the larger sums are nearest the truth; indeed, the grants and charters still in existence, or noticed by historians, would rather lead to the supposition that the revenues must have been even greater. Parsimony in the endowment of religious buildings, was not a prevailing vice in the eleventh and twelfth centuries. Least of all was it likely that it should be practised in the case of establishments, thus founded in expiation of the transgressions of wealthy and powerful sinners. Page after page, in the charters, is filled with the list of those, who, with

"Lands and livings, many a rood,
Had gifted the shrine for their soul 's repose."

The privileges and immunities enjoyed by these abbeys were very extensive. Both of them were from their origin exempted by Pope Alexander IInd, with the consent of Odo, Bishop of Bayeux, from all episcopal jurisdiction; and both had full power, as well spiritual as ecclesiastical, over the members of their own communities, and over the parishes dependent upon them; with no other appeal than to the archbishop of Rouen, or to the Pope. Express permission was likewise given to the abbot of St. Stephen's, by virtue of a bull from Pope Clement VIIth, to wear a gold mitre studded with precious stones, and a ring and sandals, and other episcopal ornaments.

Many of the monuments and deeds of the greater abbey are now in the prefecture of the department. The original chartulary or register was saved by the Abbé de la Rue, and is at this time preserved in his valuable collection. The charters of the Trinity were hid, during the revolution, by the nuns, who secreted them beneath the tiling of a barn. They were discovered there not long since; but damp and vermin had rendered them wholly illegible.

Lanfranc, whose services at Rome well deserved every distinction that his sovereign could bestow, was the first abbot of St. Stephen's. Upon his translation to the see of Canterbury, he was succeeded by William, who was likewise subsequently honored with an archiepiscopal mitre. The third abbot, Gislebert, was bishop of Evreux; and, though the series was not continued through an uninterrupted line of equal dignity, the office of abbot of this convent was seldom conferred, except upon an individual of exalted birth. Eight cardinals, two of them of the noble houses of Medici and Farnese, and three others, still more illustrious, the cardinals Richelieu, Mazarine, and Fleury, are included in the list, though in later times the abbacy was held in commendam by these powerful prelates, whilst all the internal management of the house devolved upon a prior. Amongst the abbots will also be found Hugh de Coilly, grandson of King Stephen, Anthony of Bourbon, a natural son of Henry IVth of France, and Charles of Orléans, who was likewise of royal extraction.—St. Stephen was selected as the patron of the abbey, in consequence of the founder having bestowed upon it the head of the protomartyr, together with one of his arms, and a phial of his blood, and the stone with which he was killed.

Lyceum & Church of St. Stephen at Caen.

The monastic buildings now serve for what, in the language of revolutionary and imperial France, was called a Lycée, but which has since assumed the less heathen appellation of a college. They constitute a fine edifice, and, seen from a short distance, in conjunction with the east end of the church, they form a grand tout-ensemble. The abbey church, from this point of view, has somewhat of an oriental character: the wide sweep of the semi-circular apsis, and the slender turrets and pyramids that rise from every part of the building, recal the idea of a Mahometan mosque. But the west end is still more striking than the east; and if, in the interior of the church of the Trinity, we had occasion to admire the beautiful quality of the Caen stone, our admiration of it was more forcibly excited here: notwithstanding the continual exposure to wind and weather, no part appears corroded, or discolored, or injured. A character of magnificence, arising in a great measure from the grand scale upon which it is built, pervades this front. But, to be regarded with advantage, it must be viewed as a whole: the parts, taken separately, are unequal and ill assorted. The simplicity of the main division approaches to meanness. Its three door-ways and double tier of windows appear disproportionally small, when contrasted with the expanse of blank wall; and their returns are remarkably shallow. The windows have no mouldings whatever, and the pillars and archivolts of the doors are very meagre. The front consists of three compartments, separated by flat buttresses; the lateral divisions rising into lofty towers, capped with octagon spires. The towers are much ornamented: three tiers of semi-circular arches surround the upper divisions; the arches of the first tier have no mouldings or pillars; the upper vary in pattern, and are enriched with pillars and bands, and some are pierced into windows.—Twelve pinnacles equally full of arches, some pointed, others semi-circular, surround each spire. Similar pinnacles rise from the ends of the transepts and the choir.—The central tower, which is short and terminates in a conical roof, was ruined by the Huguenots, who undermined it, thinking that its fall would destroy the whole building. Fortunately, however, it only damaged a portion of the eastern end; the reparations done to which have occasioned a discrepancy of style, that is injurious to the general effect. But the choir and apsis were previously of a different æra from the rest of the edifice. They were raised by the Abbot Simon de Trevieres, in the beginning of the fourteenth century.—I am greatly mistaken, if a real Norman church ever extended farther eastward than the choir.

The building is now undergoing a thorough repair, at the expence of the town. No other revenues, at present, belong to it, except the *sous* which are paid for chairs during mass.

A friend, who is travelling through Normandy, describes the interior in the following manner; and, as I agree with him in his ideas, I shall borrow his description:—"Without doubt, the architect was conversant with Roman buildings, though he has Normanized their features, and adopted the lines of the basilica to a barbaric temple. The Coliseum furnished the elevation of the nave;—semi-circular arches surmounted by another tier of equal span, and springing at nearly an equal height from the basis of the supporting pillars. The architraves connecting the lower rows of pillars are distinctly enounced. The arches which rise from them have plain bold mouldings. The piers between each arch are of considerable width. In the centre of each pier is a column, which ascends as usual to the vault. These columns are alternately simple and compound. The latter are square pilasters, each fronted by a cylindrical column, which of course projects farther into the nave than the simple columns; and thus the nave is divided into bays. This system is imitated in the gothic cathedral, at Sens. The square pilaster ceases at about four-fifths of its height: then two cylindrical pillars rise from it, so that, from that point, the column becomes clustered. Angular brackets, sculptured with knots, grotesque heads, and foliage, are affixed to the base of these derivative pillars. A bold double-billeted moulding is continued below the clerestory, whose windows adapt themselves to the binary arrangement of the bays. A taller arch is flanked by a smaller one on the right or the left side, as its situation requires. These are supported by short massy pillars: an embattled moulding runs round the windows.

"In the choir the arches become pointed, but with Norman mouldings: the apsis is a re-construction. In that portion of the choir, which seems original, there are pointed windows formed by the interlacing of circular arches: these light the gallery.

"The effect produced by the perspective of the interior is lofty and palatial. The ancient masonry of the exterior is worthy of notice. The stones are all small, perhaps not exceeding nine or twelve inches: the joints are about three-quarters of an inch."

At the north-west angle of the nave has been built a large chapel, comparatively a modern erection; and in the centre of this lies Matilda's gravestone.—There is no other chapel to the nave, and, as usual, no monument in any portion of the church; but in front of the high altar is still to be seen the flat stone, placed there in 1742, in memory of the Conqueror, and bearing the epitaph—

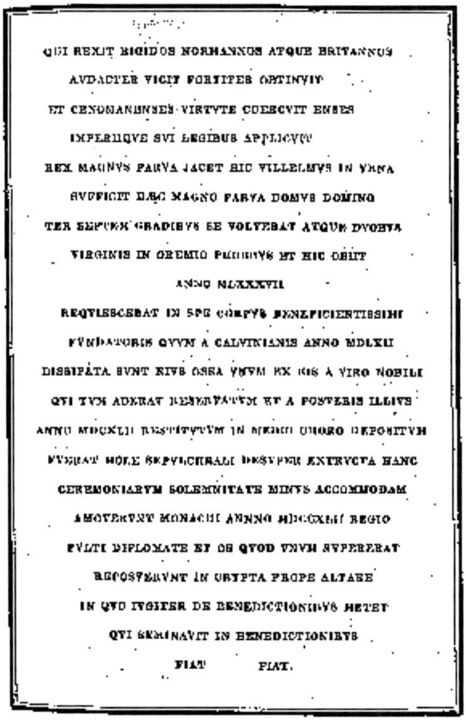
QVI REXIT RIGIDOS NORMANNOS ATQUE BRITANNOS
AVDACTER VICIT FORTITER OBTINVIT
ET CENOMANENSES VIRTVTE COERCVIT ENSES
IMPERIIQVE SVI LEGIBUS APPLICVIT
REX MAGNVS PARVA JACET HIC GVILLELMVS IN VRNA
SVFFICIT HÆC MAGNO PARVA DOMVS DOMINO
TER SEPTEM GRADIBVS SE VOLVERAT ATQVE DVOBVS
VIRGINIS IN GREMIO PHOEBVS ET HIC OBIIT
ANNO MLXXXVII
REQVIESCEBAT IN SPE CORPVS BENEFICENTISSIMI
FVNDATORIS QVVM A CALVINIANIS ANNO MDLXII
DISSIPATA SVNT EIVS OSSA VNVM EX IIS A VIRO NOBILI
QVI TVM ADERAT RESERVATVM ET A POSTERIS ILLIVS
ANNO MDCXLII RESTITVTVM IN MEDIO CHORO DEPOSITVM
FVERAT MOLE SEPVLCHRALI DESVPER EXTRVCTA HANC
CEREMONIARVM SOLEMNITATE MINVS ACCOMMODAM
AMOVERVNT MONACHI ANNO MDCCXLII REGIO
FVLTI DIPLOMATE ET OB QVOD VNVM SVPERERAT
REPOSVERVNT IN CRYPTA PROPE ALTARE
IN QVO IVGITER DE BENEDICTIONIBVS METET
QVI SEMINAVIT IN BENEDICTIONIBVS
FIAT FIAT.

QUI REXIT RIGIDOS NORMANNOS ATQUE BRITANNOS
AVDACTER VICIT FORTITER OBTINVIT
ET CENOMANENSES VIRTVTE COERCVIT ENSES
IMPERIIQVE SVI LEGIBUS APPLICVIT

REX MAGNVS PARVA JACET HIC VILLELMVS IN VRNA
SVFFICIT HÆC MAGNO PARVA DOMVS DOMINO
TER SEPTEM GRADIBVS SE VOLVERAT ATQUE DVOBVS
VIRGINIS IN GREMIO PHOEBVS ET HIC OBIIT
ANNO MLXXXVII
REQVIESCEBAT IN SPE CORPVS BENEFICIENTISSIMI
FVNDATORIS QVVM A CALVINIANIS ANNO MDLXII
DISSIPATA SVNT EIVS OSSA VNVM EX EIS A VIRO NOBILI
QVI TVM ADERAT RESERVATVM ET A POSTERIS ILLIVS
ANNO MDCXLII RESTITVTVM IN MEDIO CHORO DEPOSITVM
FVERAT MOLE SEPVLCHRALI DESVPER EXTRVCTA HANC
CEREMONIARVM SOLEMNITATE MINVS ACCOMMODAM
AMOVERVNT MONACHI ANNO MDCCXLII REGIO
FVLTI DIPLOMATE ET OS QVOD VNVM SVPERERAT
REPOSVERVNT IN CRYPTA PROPE ALTARE
IN QVO IVGITER DE BENEDICTIONIBVS METET
QVI SEMINAVIT IN BENEDICTIONIBVS
FIAT FIAT

The poetical part of this epitaph was composed by Thomas, archbishop of York, and was engraved upon the original monument, as well as upon a plate of gilt copper, which was found within the sepulchre when it was first opened. Many other poets, we are told by Ordericus Vitalis, exercised their talents upon the occasion; but none of their productions were deemed worthy to be inscribed upon the tomb. The account of the opening of the vault is related by De Bourgueville, from whom it has been already copied by Ducarel; but the circumstances are so curious, that I shall offer no apology for telling a twice-told tale. From Ordericus Vitalis also we may borrow some details respecting the funeral of the Conqueror, which, though strictly appertaining to English history, have never yet, I believe, appeared in an English dress.

In speaking of the church of St. Gervais at Rouen, I have already briefly alluded to the melancholy circumstances by which the death of this monarch was attended. The sequel of the story is not less memorable.

The king's decease was the signal for general consternation throughout the metropolis of Normandy. The citizens, panic struck, ran to and fro, as if intoxicated, or as if the town were upon the point of being taken by assault. Each asked counsel of his neighbor, and each anxiously turned his thoughts to the concealing of his property. When the alarm had in some measure subsided the monks and clergy made a solemn procession to the abbey of St. Georges, where they offered their prayers for the repose of the soul of the departed Duke; and archbishop William commanded that the body should be carried to Caen, to be interred in the church of St. Stephen, which William had founded. But the lifeless king was now deserted by all who had participated in his munificence and bounty. Every one of his brethren and relations had left him; nor was there even a servant to be found to perform the last offices to his departed lord. The care of the obsequies was finally undertaken by Herluin, a knight of that district, who, moved by the love of God and the honor of his nation, provided at his own expence, embalmers, and bearers, and a hearse, and conveyed the corpse to the Seine, whence it was carried by land and water to the place of its destination.

Upon the arrival of the funeral train at Caen, it was met by Gislebert, bishop of Evreux, then abbot of St. Stephen's, at the head of his monks, attended with a numerous throng of clergy and laity; but scarcely had the bier been brought within the gates, when the report was spread that a dreadful fire had broken out in another part of the town, and the Duke's remains were a second time deserted. The monks alone remained; and, fearful and irresolute, they bore their founder "with candle, with book, and with knell," to his last home. Ordericus Vitalis enumerates the principal prelates and barons assembled upon this occasion; but he makes no mention of the Conqueror's son, Henry, who, according to William of Jumieges, was the only one of the family that attended, and was also the only one worthy of succeeding to such a father.—Mass had now been performed, and the body was about to be committed to the ground, "ashes to ashes, dust to dust," when, previously to this closing part of the ceremony, Gislebert mounted the pulpit, and delivered an oration in honor of the deceased.—He praised his valor, which had so widely extended the limits of the Norman dominion; his ability, which had elevated the nation to the highest pitch of glory; his equity in the administration of justice; his firmness in correcting abuses; and his liberality towards the monks and clergy; then, finally, addressing the people, he besought them to intercede with the Almighty for the soul of their prince, and to pardon whatsoever transgression he might have been guilty of towards any of them.—At this moment, one Asselin, an obscure individual, starting from the crowd, exclaimed with a loud voice, "the ground upon which you are standing, was the site of my father's dwelling. This man, for whom you ask our prayers, took it by force from my parent; by violence he seized, by violence he retained it; and, contrary to all law and justice, he built upon it this church, where we are assembled. Publicly, therefore, in the sight of God and man, do I claim my inheritance, and protest against the body of the plunderer being covered with my turf."—The appeal was attended with instant effect; bishops and nobles united in their entreaties to Asselin; they admitted the justice of his claim; they pacified him; they paid him sixty shillings on the spot by way of recompence for the place of sepulture; and, finally, they satisfied him for the rest of the land.

But the remarkable incidents doomed to attend upon this burial, were not yet at an end; for at the time when they were laying the corpse in the sarcophagus, and were bending it with some force, which they were compelled to do, in consequence of the coffin having been made too short, the body, which was extremely corpulent, burst, and so intolerable a stench issued from the grave, that all the perfumes which arose from all the censers of the priests and acolytes were of no avail; and the rites were concluded in haste, and the assembly, struck with horror, returned to their homes.

The latter part of this story accords but ill with what De Bourgueville relates. We learn from this author, that four hundred and thirty years subsequent to the death of the Conqueror, a Roman cardinal, attended by an archbishop and bishop, visited the town of Caen, and that his eminence having expressed a wish to see the body of the duke, the monks yielded to his curiosity, and the tomb was opened, and the corpse discovered in so perfect a state, that the cardinal caused a portrait to be taken from the lifeless features.—It is not worth while now to inquire into the truth of this story, or the fidelity of the resemblance. The painting has disappeared in the course of time: it hung for a while against the walls of the church, opposite to the monument; but it was stolen during the tumults caused by the Huguenots, and was broken into two pieces, in which state De Bourgueville saw it a few years afterwards, in the hands of a Calvinist, one Peter Hodé, the gaoler at Caen, who used it in the double capacity of a table and a door.—The worthy magistrate states, that he kept the picture, "because the abbey-church was demolished."

He was himself present at the second violation of the royal tomb, in 1572; and he gives a piteous account of the transaction. The monument raised to the memory of the Conqueror, by his son, William Rufus, under the superintendance of Lanfranc, was a production of much costly and elaborate workmanship: the shrine, which was placed upon the mausoleum, glittered with gold and silver and precious stones. To complete the whole, the effigy of the king had been added to the tomb, at some period subsequent to its original erection.—A monument like this naturally excited the rapacity of a lawless banditti, unrestrained by civil or military force, and inveterate against every thing that might be regarded as connected with the Catholic worship.—The Calvinists were masters of Caen, and, incited by the information of what had taken place at Rouen, they resolved to repeat the same outrages. Under the specious pretext of abolishing idolatrous worship, they pillaged and ransacked every church and monastery: they broke the painted windows and organs, destroyed the images, stole the ecclesiastical ornaments, sold the shrines, committed pulpits, chests, books, and whatever was combustible, to the fire; and finally, after having wreaked their vengeance upon eyery thing that could be made the object of it, they went boldly to the town-hall to demand the wages for their labors.—In the course of these outrages the tomb of the Conqueror at one abbey, and that of Matilda at the other, were demolished. And this was not enough; but a few days afterwards, the same band returned, allured by the hopes of farther plunder. It was customary in ancient times to deposit treasures of various kinds in the tombs of sovereigns, as if the feelings of the living passed into the next stage of existence;—

"... quæ gratia currûm
Armorumque fuit vivis, quæ cura nitentes
Pascere equos, eadem sequitur tellure repostos."

The bees that adorned the imperial mantle of Napoléon were found in the tomb of Childeric. A similar expectation excited the Huguenots, at Caen. They dug up the coffin: the hollow stone rung to the strokes of their daggers: the vibration proved that it was not filled by the corpse; and nothing more was wanted to seal its destruction.

De Bourgueville, who went to the spot and exerted his eloquence to check this last act of violence, witnessed the opening of the coffin. It contained the bones of the king, wrapped up in red taffety, and still in tolerable preservation; but nothing else. He collected them, with care, and consigned them to one of the monks of the abbeys who kept them in his chamber, till the Admiral de Châtillon entered Caen at the head of his mercenaries, on which occasion the whole abbey was plundered, and the monks put to flight, and the bones lost. "Sad doings, these," says De Bourgueville, *"et bien peu réformez!"*—He adds, that one of the thigh-bones was preserved by the Viscount of Falaise, who was there with him, and begged it from the rioters, and that this bone was longer by four fingers' breadth than that of a tall man. The bone thus preserved, was re-interred, after the cessation of the troubles: it is the same that is alluded to in the inscription, which also informs us that a monument was raised over it in 1642, but was removed in 1742, it being then considered as an incumbrance in the choir.

With this detail I close my letter. The melancholy end of the Conqueror, the strange occurrences at his interment, the violation of his grave, the dispersion of his remains, and the demolition and final removal of his monument, are circumstances calculated to excite melancholy emotions in the mind of every one, whatever his condition in life. In all these events, the religious man traces the hand of retributive justice; the philosopher regards the nullity of sublunary grandeur; the historian finds matter for serious reflection; the poet for affecting narrative; the moralist for his tale; and the school-boy for his theme.—Ordericus Vitalis sums the whole up admirably. I should spoil his language were I to attempt to translate it; I give it you, therefore, in his own words:—"Non fictilem tragoediam venundo, non loquaci comoedia cachinnantibus parasitis faveo: sed studiosis lectoribus varios eventus veraciter intimo. Inter prospera patuerunt adversa, ut terrerentur terrigenarum corda. Rex quondam potens et bellicosus, multisque populis per plures Provincias metuendus, in area jacuit nudus, et a suis, quos genuerat vel aluerat, destitutus. Aere alieno in funebri cultu indiguit, ope gregarii pro sandapila et vespilionibus conducendis eguit, qui tot hactenus et superfluis opibus nimis abundavit. Secus incendium a formidolosis vectus est ad Basilicam, liberoque solo, qui tot urbibus et oppidis et vicis principatus est, caruit ad sepulturam. Arvina ventris ejus tot delectamentis enutrita cum dedecore patuit, et prudentes ac infrunitos, qualis sit gloria carnis, edocuit[79]."

Footnotes:

[76] Anglo-Norman Antiquities, p. 45.

[77] See Cotman's Architectural Antiquities of Normandy, t. 24-33.

[78] A detailed account of the proceedings on this occasion, is given in the Journal Politique du Département du Calvados, for March 21, and May 6, 1819.—The first attempt at the discovery of Matilda's coffin, was made in March, 1818, and was confined to the chapter-house: the matter then slept till the following March, when Count de Montlivault, attended by the Bishop of Bayeux, Mr. Spencer Smythe, and other gentlemen, prosecuted his inquiries within the

church itself, and, immediately under the spot where her monument stood, discovered a stone coffin, five feet four inches long, by eleven inches deep, and varying in width from twenty inches to eleven. Within this coffin was a leaden box, soldered down; and, in addition to the box, the head of an effigy of a monk, in stone, and a portion of a skull-bone filled with aromatic herbs, and covered with a yellowish-white membrane, which proved, upon examination, to be the remains of a linen cloth. The box contained various bones, that had belonged to a person of nearly the same height as Matilda is described to have been. No doubt seemed to remain but that the desideratum was discovered. The whole was therefore carefully replaced; and the prefect ordered that a new tomb should be raised, similar to that which was destroyed at the revolution; and that the slab, with the original epitaph, should be laid on the top; that copies of the former inscription, stating how the queen's remains had been re-interred by the abbess, in 1707, should be added to two of the sides; that to the third should be affixed the ducal arms of Normandy; and that the fourth should bear the following inscription:—

"Ce tombeau renfermant les dépouilles mortelles

de l 'illustre Fondatrice de cette Abbaye,

renversé pendant les discordes civiles,

et déplacé depuis une longue série d 'années,

a été restauré, conformément au voeu des

amis de la religion, de l 'antiquité et des arts,

1819.

Casimir, comte de Montlivault, conseiller d 'état, préfet.

Léchaudé d 'Anisy, directeur de l 'Hospice."

The ceremony of the re-interment was performed with great pomp on the fifth of May; and the Bishop of Bayeux pronounced a speech on the occasion, that does him credit for its good sense and affecting eloquence.

[79] Hist. Normannorum Scriptores, p. 662.

LETTER XXVI
PALACE OF THE CONQUEROR—HERALDIC TILES—PORTRAITS OF WILLIAM AND MATILDA—MUSEUM—PUBLIC LIBRARY—UNIVERSITY—ACADEMY—EMINENT MEN—HISTORY OF CAEN

(*Caen, August,* 1818.)

Within the precincts of the abbey of St. Stephen are some buildings, which do not appear to have been used for monastic purposes. It is supposed that they were erected by William the Conqueror, and they are yet called his palace. Only sixty years ago, when Ducarel visited Caen, these remains still preserved their original character.

He describes the great guard-chamber and the barons' hall, as making a noble appearance, and as being perhaps equally worth the notice of an English antiquary as any object within the province of Normandy. The walls of these rooms are standing, but dilapidated and degraded; and they have lost their architectural character, which, supposing Ducarel's plate to be a faithful representation, must have been very decisive. It is scarcely possible to conceive how any man, with such a specimen of the palace before his eyes, could dream of its being coeval with the Norman conquest: every portion is of the pointed style, and even of a period when that style was no longer in its purity. Possibly, indeed, other parts of the edifice may have been more ancient; such certainly was the "Conqueror's kitchen," a singular octagon building, with four tall slender chimneys capped with perforated cones. This was destroyed many years ago; but Ducarel obtained an original drawing of it, which he has engraved. Amongst the ruins there is a chimney which perhaps belonged to this building.—The guard-chamber and barons' hall are noble rooms: the former is one hundred and ninety feet in length and ninety in breadth. You remember how admirably the Lay of the Last Minstrel opens with a description of such a hall, filled with knights, and squires, and pages, and all the accompaniments of feudal state. I tried, while standing by these walls, to conjure up the same pictures to my imagination, but it was impossible; so desolate and altered was every thing around, and so effectually was the place of baronial assemblage converted into a granary. The ample fire-place still remains; but, cold and cheerless, it looks as if had been left in mockery of departed splendor and hospitality. I annex a sketch of it, in which you will also see a few scattered tiles, relics of the magnificent pavement that once covered the floor.

Fire-place in the Ducal palace at Caen.

This pavement has been the subject of much learned discussion; because, if the antiquity of the emblazoned tiles could be established, (which it certainly cannot) we should then have a decisive proof of the use of armorial bearings in the eleventh century. Nearly the whole of these tiles are now removed. After the abbey was sold, the workmen entirely destroyed the tiles, breaking them with their pick-axes. The Abbé de la Rue, however, collected an entire set of them; and others have been preserved by M. Lair, an antiquary of Caen.—Ducarel thus describes the pavement when perfect: "The floor is laid with tiles, each near five inches square, baked almost to vitrification. Eight rows of these tiles, running from east to west, are charged with different coats of arms, said to be those of the families who attended Duke William in his invasion of England. The intervals between each of these rows are filled up with a kind of tessellated pavement, the middle whereof represents a maze or labyrinth, about ten feet in diameter, and so artfully contrived that were we to suppose a man following all the intricate meanders of its volutes, he could not travel less than a mile before he got from one end to the other. The remainder of the floor is inlaid with small squares of different colors, placed alternately, and formed into draught or chess-boards, for the amusement of the soldiers while on guard."

Such is the general description of the floors of this apartment: with regard to the date of the tiles, Ducarel proceeds to state that "it is most probable the pavement was laid down in the latter part of the reign of King John, when he was loitering away his life at Caen, with the beautiful Isabel of Angoulême, his queen; during which period, the custom of wearing coats of arms was introduced."—Common tradition assigns the tiles to higher date, making them coeval with the conquest; and this opinion has not been without supporters. It was strenuously defended by Mr. Henniker Major, who, in the year 1794, printed for private distribution, two letters upon the subject, addressed to Lord Leicester, in which he maintained this opinion with zeal and laborious research. To the letters were annexed engravings of twenty coats of arms, the whole, as he observes, that were represented on the pavement; for though the number of emblazoned tiles was considerable, the rest were all repetitions[80]. The same observation was found in the inscription attached to a number of the tiles, which the monks kept framed for public inspection, in a conspicuous part of the monastery; and yet some of the armorial bearings in this very selection, differ from any of those figured by Mr. Henniker Major. The Abbé de la Rue has also many which are not included in Mr. Henniker Major's engravings. In one of the coats the arms are quartered, a practice that was not introduced till the reign of Edward IIIrd. The same quarterings are also found upon an escutcheon, placed over the door that leads to the apartment. This door is a flattened arch, with an ogee canopy, the workmanship probably of the fourteenth century.

To the same date I should also refer the tiles; and possibly the whole palace was built at that period. There are no records of its erection; no document connects its existence with the history of the duchy; no author relates its having been suffered to fall into decay. So striking an absence of all proof, and this upon a point where evidence of different kinds might naturally have been expected, may warrant a suspicion how far the building was ever a royal palace, according to the strict import of the town. A friend of mine supposes that these buildings may have been the king's lodgings. During the middle ages it was usual for monarchs in their progresses, to put up at the great abbeys; and this portion of the convent of St. Stephen may have been intended for the accommodation of the royal guests.

The assigning of a comparatively modern date to the pavement, does not necessarily interfere with the question as to the antiquity of heraldic bearings. The coats of arms which are painted upon the tiles may have been designed to represent those of the nobility who attended Duke William on his expedition to England: it is equally possible that they embraced a more general object, and were those of the principal families of the duchy—De Thou gives his suffrage in favor of the former opinion, but Huet of the latter; and the testimony of the bishop must be allowed, in this case, to outweigh that of the president.—Huet also says, that it is matter of notoriety that the tiles were laid down towards the close of the fourteenth century. He mentions, however, no authority for the assertion; and less credit perhaps will be given to it than it deserves, from his having stated just before, that the abbey and palace were contemporary structures.

Upon the outside wall of a chapel that is supposed to have belonged to the same palace, were ancient fresco paintings of William and Matilda, and of their sons, Robert and William Rufus. They are engraved by Montfaucon[81], and are supposed by him, probably with reason, to be coeval with the personages they represent. The figures are standing upon animals, the distribution of which is the most remarkable circumstance connected with the portraits. To the king is assigned a dog; to the queen a lion: the eldest son has the same symbol as his father; the younger rests upon a two-bodied beast, half swine, half bird, the bodies uniting in a female head.—Upon the same plate, Montfaucon has given a second whole-length picture of the conqueror, which represents him with the crown upon his head, and the sceptre in his hand. Considering the costume, he observes with justice that it cannot have been painted earlier than the latter part of the fourteenth century. Ducarel, who, as usual, has copied the Benedictine's engravings, says that, in his time, the same portrait existed in fresco over a chimney-piece in the porter's lodge.—We saw two copies of it; the one in the sacristy of the abbey church, the other in the museum, an establishment which may, without injustice to the honors of Caen, be dismissed with the brief observation, that, though three rooms are appropriated to the purpose, there is a very scanty assortment of pictures, and their quality is altogether ordinary.

The public library is a handsome apartment, one hundred and thirty feet in length, and it contains about twenty thousand volumes, mostly in good condition; but a great proportion of the books are of a description little read, being old divinity. To the students of the university, this establishment is of essential service; and on this account it is to be regretted, that the very scanty revenue with which it is endowed, amounting only to twelve hundred francs per annum, prevents the possibility of any material increase to the collection, except in the case of such books as the liberality of the state contributes. And these are principally works of luxury and great expence, which might advantageously be exchanged for the less costly productions of more extensive utility. We inquired in vain after manuscripts and specimens of early typography. None were to be found; and yet they might surely have been expected here; for a public library has existed in Caen from an early part of the last century, and, previous to the revolution, it was enriched with various donations. M. de Colleville presented to it the whole of the collection of the celebrated Bochart; Cavelier, printer to the university, a man known by several treatises on Roman antiquities, added a donation of two thousand volumes; and Cardinal de Fleury, who considered it under his especial protection, gave various sums of money for the purchase of books, and likewise provided a salary for the librarian. I suspect that no small proportion of the more valuable volumes, have been dispersed or stolen. Round the apartment hang portraits of the most eminent men of Caen: tablets are also suspended, for the purpose of commemorating those who have been benefactors to the library; but the tablets at present are blank.

For its university Caen is indebted to Henry VIth, who, anxious to give éclat and popularity to British rule, founded a college by letters patent, dated from Rouen, in January, 1431. The original charter restricted the objects of the university to education in the canon and civil law; but, five years subsequently, the same king issued a fresh patent, adding the faculties of theology and the arts; and, in the following year, he still farther added the faculty of medicine.—To give permanency to the work thus happily begun, the states of Normandy preferred their petition to Pope Eugene IVth, who issued two bulls, dated the thirtieth of May, 1437, and the nineteenth of May, 1439, by which the new university received the sanction of the holy see, and was placed upon the same footing as the other universities of the kingdom. The Bishop of Bayeux was at the same time appointed chancellor; and sundry apostolical privileges were conceded, which have been confirmed by subsequent pontiffs.—Thus Normandy, as is admitted by De Bourgueville, owed good as well as evil to her English sovereigns; but Charles VIIth had no sooner succeeded in expelling our countrymen from the province, than jealousy arose in his breast, at finding them in possession of such a title to the gratitude of the people, and he resolved to run the risk of destroying what had been done, rather than lose the opportunity of gratifying his personal feeling. The university was therefore dissolved in 1450, that a new one might hereafter be founded by the new sovereign. The king thought it necessary to vary in some degree from the example of his predecessor; and for this purpose he had recourse to the extraordinary expedient of abolishing the faculty of law. A petition, however, from the states, induced him to replace the whole upon its original footing in 1452, and it continued till the time of the revolution to have all the five faculties, and to be the only one in France that retained them. Two years only intervened between the dates of the patents issued by Charles VIIth, upon the subject of this university; yet there is a remarkable difference in their language. The first of them, which is obviously intended to disparage Caen, styles it a large town, scantily inhabited, without manufactures or commerce, and destitute of any great river to afford facilities towards the transport of the produce of the country. The second was designed to have an opposite tendency; and in this, the people of Caen are praised for their acuteness, and the town for its excellent harbor and great rivers. The patent also adds, that the nearest university, that of Paris, is fifty leagues distant.

In the estimation, at least, of the inhabitants, the university of Caen ranks at present the third in France; Paris and Strasbourg being alone entitled to stand before it. The faculty of law retains its old reputation, and the legal students are quite the pride of the university. Since the peace, many young jurisprudents from Jersey and Guernsey have resorted to it. Medical students generally complete their education at Paris, where it is commonly considered in France, that, both in theory and practice, the various branches of this faculty have nearly attained the acmè of perfection. The students, who amount to just five hundred, are under the care of twenty-six professors, many of them men of distinguished talents. The Abbé de la Rue fills the chair of history; M. Lamouroux, that of the natural sciences. They receive their salaries wholly from the government; their emoluments continue the same, whether the students crowd to hear their courses, or whether they lecture to empty benches. It is strictly forbidden to a student to attempt to make any remuneration to a professor, or even to offer him a present of any kind. The whole of the dues paid by the scholars go to the state; and the state in its turn, defrays the expences of the establishment.

There is likewise at Caen an Academy of Sciences, Arts and Belles Lettres, which has published two volumes; not, strictly speaking, of its Transactions, but exhibiting a brief outline of the principal papers that have been read at the meetings. The antiquarian dissertations of the Abbé de la Rue, which they contain, are of great merit; and it is much to be regretted, that they have not appeared in a more extended form. A chartered academy was first founded here in the year 1705; and it continued to exist, till it was suppressed, like all others throughout France, at the revolution. The present establishment arose in 1800, under the auspices of General Dugua, then prefect of the department, who had been urged to the task by the celebrated Chaptal, Minister of the Interior.—Some interesting, letters are annexed to the second part of the poems of Mosant de Brieux, in which, among much curious information relative to Caen, he describes the literary meetings that led to the foundation of the first academy. The town at that time could boast an unusual proportion of men of talents. Bochart, author of *Sacred Geography*; Graindorge, who had published *De Principiis Generationîs*;

Huet, a man seldom mentioned, without the epithet *learned* being attached to his name; and Halley and Ménage, authors almost equally distinguished, were amongst those who were associated for the purposes of acquiring and communicating information.

Indeed, Caen appears at all times to have been fruitful in literary characters. Huet enumerates no fewer than one hundred and thirty-seven, whom he considers worthy of being recorded among the eminent men of France. The greater part of them are necessarily unknown to us in England; and allowance must be made for a man who is writing upon a subject, in which self-love may be considered as in some degree involved; the glory of our townsmen shining by reflection upon ourselves. A portion, however, of the number, are men whose claims to celebrity will not be denied.—Such, in the fifteenth century, were the poets John and Clement Marot; such was the celebrated physician, Dalechamps, to whom naturalists are indebted for the Historia Plantarum; such the laborious lexicographer, Constantin; and, not to extend the catalogue needlessly, such above all was Malherbe. The medal that has been struck at Caen in honor of this great man, at the expence of Monsieur de Lair, bears for its epigraph, the three first words of Boileau 's eulogium—"Enfin Malherbe vint."—The same inscription is also to be seen upon the walls of the library. So expressive a beginning prepares the reader for a corresponding sequel; and I should be guilty of injustice towards this eminent writer, were I not to quote to you the passage at length.—

"Enfin, Malherbe vint, et le premier en France
Fit sentir dans les vers une juste cadence:
D 'un mot mis en sa place enseigna le pouvoir,
Et reduisit la muse aux règles du devoir.
Par ce sage écrivain, la langue repareé,
N 'offrit plus rien de rude à l 'oreille épureé.
Les stances avec grâce apprirent à tomber,
Et le Vers sur le Vers n 'osa plus enjamber."

Wace and Baudius, though not born at Caen, have contributed to its honor, by their residence here. Baudius was appointed to the professorship of law in the university, by the President de Thou; but he disagreed with his colleagues, and soon removed to Leyden, where he filled the chair of history till his death. Some of his earlier letters, in the collection published by Elzevir, are dated from Caen. His Iambi, directed against his brethren of this university, are scarcely to be exceeded for severity, by the bitterest specimens of a style proverbially bitter. Their excessive virulence defeated the writer's aim; but there is an elegance in the Latinity of Baudius, and a degree of feeling in his sentiments, which will ensure a permanent existence to his compositions, and especially to his poems.—He it was who called forth the severe saying of Bayle, that "many men of learning render themselves contemptible in the places where they live, while they are admired where they are known only by their writings."—Wace was a native of Jersey, but an author only at Caen. The most celebrated of his works is Le Roman de Rou et des Normans, written in French verse. He dedicated this romance to our Henry IInd, who rewarded him with a stall in the cathedral at Bayeux.

Quitting the departed for the living, I send you a profile of M. Lamouroux, the professor of natural history at this university, to whom we have been personally indebted for the kindest attention. His name is well known to you, as that of a man who has, perhaps, deserved more than any other individual at the hands of every student of marine Botany. His treatises upon the Classification of the Submersed Algæ, have been honored with admission in the Mémoires du Muséum d'Histoire Naturelle, and have procured him the distinction of being elected into the National Institute: his subsequent publication on the Corallines, is an admirable manual, in a very difficult branch of natural history; and he is now preparing for the press, a work of still greater labor and more extensive utility, an arrangement of the organized fossils found in the vicinity of Caen.

The whole of this neighborhood abounds in remains of the antediluvian world: they are found not only in considerable quantity, but in great perfection. In the course of last year; a fossil crocodile was dug up at Allemagne, a village about a mile distant, imbedded in blue lias. Other specimens of the same genus, comprising, as it appears, two species, both of them distinct from any that are known in a living state, had previously been discovered in a bed of similar hard blue limestone, near Havre and Honfleur, as well as upon the opposite shores of England. But the Caen specimen is the most interesting of any, as the first that has been seen with its scales perfect; and the naturalists here have availed themselves of the opportunity thus afforded them, to determine it by a specific character, and give it the name of *Crocodilus Cadomensis*.

The civil and ecclesiastical history of Caen will be amply illustrated in the forthcoming volumes of the Abbé de la Rue, as he is preparing a work on the subject, à l'instar of the Essays of St. Foix. In the leading events of the duchy, we find the town of Caen had but little share. It is only upon the occasion of two sieges from our countrymen, the one in 1346, the other in 1417, that it appears to have acted a prominent part. The details of the first siege are given at some length by Froissart.—Edward IIIrd, accompanied by the Black Prince, had landed at La Hogue; and, meeting with no effectual resistance, had pillaged the towns of Barfleur, Cherbourg, Carentan, and St. Lô, after which he led his army hither. Caen, as Froissait tells us, was at that time "large, strong, and full of drapery and all other sorts of merchandize, rich citizens, noble dames and damsels, and fine churches." In its defence were assembled the Constable of France, with the Counts of Eu, Guignes, and Tancarville. But the wisdom of the generals was defeated by the impetuosity of the citizens. They saw themselves equal in number to the invaders, and, without reflecting how little numerical superiority avails in war against experience and tactics, they required to be led against the foe. They were so, and were defeated. The conquerors and conquered entered the city pell-mell; and Edward, enraged at the citizens for shooting upon his troops from the windows, issued orders that the inhabitants should be put to the sword, and the town burned. The mandate, however, was not executed: Sir Godfrey de Harcourt, with wise remonstrances, assuaged the anger of the sovereign, and diverted him from his purpose.—Immense were the riches taken on the occasion. The English fleet returned home loaded with cloth, and jewels, and gold, and silver plate, together with sixty knights, and upwards of three hundred able men, prisoners. This gallant exploit was shortly afterwards followed by the decisive battle of Crécy.

Caen suffered still more severely upon the occasion of its second capture; when Henry IVth marched upon the town immediately after landing at Touques. The siege was longer, and the place, taken by assault, was given up to indiscriminate plunder. Even the churches were not spared: that of the Holy Sepulchre was demolished, and, among its other treasures, a crucifix was carried away, containing a portion of the real cross, which, as we are told, testified by so many miracles its displeasure at being taken to England, that the conquerors were glad to restore it to its original destination.

From this time to the year 1450, our countrymen kept undisturbed possession of Caen. In the latter year they capitulated to the Count de Dunois, after a gallant resistance. But though the town has thenceforward remained, without interruption, subject to the crown of France, it has not therefore been always free from the miseries of warfare. A dreadful riot took place here in 1512, occasioned by the disorderly conduct of a body of six thousand German mercenaries, whom Louis XIIth introduced, by way of garrison, to guard against any sudden attack from Henry VIIIth. The character given by De Bourgueville of these Lansquenets is, that they were "drunkards who guzzle wine, cider, and beer, out of earthen pots, and then fall asleep upon the table." Three hundred lives were lost upon this occasion, on the part of the Germans alone.—In the middle of the same century, happened the civil wars, originating in the reformation: and in the course of these, Caen suffered dreadfully from the contending parties. Friend and foe conspired alike to its ruin: what was saved from the violence of the Huguenots, was taken by the treachery of the Catholics, under the plausible pretext of its being placed in security. Thus, after the Calvinists had already seized on every thing precious that fell in their way, the Duke de Bouillon, the governor of the town, commanded all the reliquaries, shrines, church-plate, and ecclesiastical ornaments, to be carried to him at the castle; and he had no sooner got them into his possession, than "all holy, rich, and precious, as they were, he caused them to be melted down, and converted into coin to pay his soldiers; and he scattered the relics, so that they have never been seen more."—Loosen but the bands of society, and you will find that, in all ages of the world, the case has been nearly the same; and, as upon the banks of the Simoeis, so upon the plains of Normandy,—

> "Seditione, dolis, scelere, atque libidine, et irâ,
>
> *Iliacos* extra muros peccatur et intra."

Footnotes:

[80] Engravings of the same tiles, and of some others, chiefly with fanciful patterns, are to be found in the Gentleman's Magazine for March 1789, LIX. p. 211, plates 2, 3. The subjects of the latter plate are those tiles which were hung in a gilt frame, on the walls of the cloister of the abbey, with an inscription, denoting whence they were taken.

[81] Monumens de la Monarchie Française, I. p. 402, t. 55.

LETTER XXVII
VIEUX—LA MALADERIE—CHESNUT TIMBER—CAEN STONE—HISTORY OF BAYEUX—TAPESTRY

(*Bayeux, August*, 1818.)

Letters just received from England oblige us to change our course entirely: their contents are of such a nature, that we could not prolong our journey with comfort or satisfaction. We must return to England; and, instead of regretting the objects which we have lost, we must rejoice that we have seen so much, and especially that we have been able to visit the cathedral and tapestry of Bayeux.

At the same time, I will not deny that we certainly could have wished to have explored the vicinity of Caen, where an ample harvest of subjects, both for the pen and pencil, is to be gathered; but the circumstances that control us would not even allow of a pilgrimage to the shrine of our Lady of la Délivrande, on the border of the English Channel, or of an excursion to the village of Vieux, in the opposite direction.—Antiquaries have been divided in opinion, concerning the nature and character of the buildings which anciently occupied the site of this village.—The remains of a Roman aqueduct are still to be seen there, and the foundations of ancient edifices are distinctly to be traced. In the course of the last century, a gymnasium was likewise discovered, of great size, constructed according to the rules laid down by Vitruvius, and a hypocaust, connected with a fine stone basin, twelve feet in diameter, surrounded by three rows of seats. Abundance of medals of the upper empire, among others, of Crispina, wife to Commodus, and Latin inscriptions and sarcophagi, are frequently dug up among its ruins[82]. Hence, a belief has commonly prevailed that during the Roman dominion in Gaul, Vieux was a city, and that Caen, which is only six miles distant, arose from its ruins. This opinion was strenuously combated by Huet; yet it subsequently found a new advocate in the Abbé Le Beuf[83]. The bishop contends that the extent of the buildings rather denotes the ruins of a fortified camp, than of a city; and he therefore considers it most probable, that Vieux was the site of an encampment, raised near the Orne, for the purpose of defending the passage of the river, at the point where it was crossed by the military road that led from the district of the Bessin, to that of the Hiesmois.—Portions of the causeway, may still be traced, constructed of the same kind of brick as the aqueduct; and the name of the village so far tends to corroborate the conjecture, that Vieux originally denoted a ford; and the word Vé, which is most probably a corruption from it, retains this signification in Norman French.—The Abbé, at the same time that he does not pretend to contradict the argument deduced from etymology, maintains that a careful comparison of the position of Vieux, with the distances marked on the Tabula Peutingeriana, and with what Ptolemy relates of certain towns adjoining the Viducassian territory, will support him in the assertion, that Vieux was the ancient Augustodurum the Viducassian capital; and that Bayeux was probably the site of Arigenus another of the towns of that tribe.—The red, veined marble of Vieux is much esteemed in France; as are also the other marbles of this department, which vary in color from a dull white, through grey, to blue. The quarries, as is generally believed, were first opened and worked by the Romans. Vieux marble is to be seen at Paris, where it was employed by Cardinal Richelieu, in the construction of the chapel of the Sorbonne.

At about a mile from Caen, on the road to Bayeux, stands the village of St. Germain de Blancherbe, more commonly called in the neighborhood la Maladerie, a name derived from the lazar-house in it, the Léproserie de Beaulieu, founded by Henry IInd, in 1161.—Robert Du Mont terms the building a wonderful work. It was a princely establishment, designed for the reception of lepers from all the parishes of Caen, except four, whose patients had an especial right to be admitted into a smaller hospital in the same place. The great hospital is now used as a house of correction. Seen from the road, it appears to be principally of modern architecture though still retaining a portion of the ancient structure; the same, probably, as is mentioned by Ducarel, who says, that "part of the magnificent chapel, which was considered as the parish church for the lepers, and ruined by the English, is turned into a large common hall for the prisoners, and separated from the other part, which is made into a chapel, by means of an iron gate, through which they may have an opportunity of hearing mass celebrated every morning."—Within the village street stands a desecrated church of the earliest Norman style, with a very perfect door-way. The present parish church, though chiefly modern, deserves attention on account of the west front, which is wholly of the semi-circular style, and is somewhat curious, from having two Norman buttresses, that rise from a string-course at the top of the basement story, (in which the arched door-way is contained,) and are thence continued upwards till they unite with the roof. The decorations round its southern entrance are also remarkable: they principally consist of a very sharp chevron moulding, interspersed with foliage and various figures.

The quarries in this village, and in that of Allemagne, on the opposite side of the Orne, supply most of the free-stone, for which Caen has, during many centuries, been celebrated. Stone of the finest quality is found in strata of different thickness, at the depth of about sixty feet below the surface of the ground. If worked much lower, it ceases to be good. It is brought up in square blocks, about nine feet wide, and two feet thick, by means of vertical wheels, placed at the mouths of the pits. When first dug from the quarry, its color is a pure and glossy white, and its texture very soft; but as it hardens it takes a browner hue, and loses its lustre.

In former days this stone was exported in great quantity to our own country. Stow, in his *Survey of London,* states that London Bridge, Westminster Abbey, and several others of our public edifices were built with it. Extracts from sundry charters relative to the quarries are quoted by Ducarel, who adds that, in his time, though many cargoes of the stone were annually conveyed by water to the different provinces of the kingdom, the exportation of it out of France was strictly prohibited, insomuch that, when it was to be sent by sea, the owner of the stone, as well as the master of the vessel on board of which it was shipped, was obliged to give security that it should not be sold to foreigners.—We omitted to inquire how far the same prohibitions still continue in force.

At but a short distance from St. Germain de Blancherbe, stands the ruined abbey of Ardennes, now the residence of a farmer; but still preserving the features of a monastic building. The convent was founded in 1138, for canons of the Præmonstratensian order. Its Celtic name denotes its antiquity, as it also tends to prove that this part of the country was covered with timber. The word, arden, signified a forest, and was thence applied, with a slight variation in orthography, to the largest forest in England, and to the more celebrated forest in the vicinity of Liege. According to tradition, the Norman ardennes consisted: of chesnut-trees. De Bourgueville tells us that timber of this description is the principal material of most of the houses in the town. John Evelyn relates the same of those in London; and in our own counties wherever a village church has been so fortunate as to preserve its ancient timber cieling, the clerk is almost sure to state that the wood is chesnut. Either this tree therefore must formerly have abounded in places where it has now almost ceased to exist, or oak timber must have been commonly mistaken for it: and we may equally adopt both these conjectures. The yew and the service, as well as the chesnut, are occasionally mentioned in old charters, and are admitted by botanists to be indigenous in England. I should doubt, however, if any one of them could now be found in a wild state; and there is a fashion in planting as well as in every thing else, which renders peculiar trees more or less abundant at different times.

About half way between Caen and Bayeux, is the village of Bretteville l'Orgueilleuse, the lofty tower of whose church, perforated with long lancet windows, and surmounted by a high spire, excites curiosity. Churches are numerous in this neighborhood, and there is no other part of Normandy, in which, architecturally considered, they are equally deserving of notice. Scarcely one is to be seen that is not marked by some peculiarity. I know not why Bretteville acquired the epithet attached to its name; and I am equally at a loss for the derivation of the word Bretteville itself; but the term must have some signification in Normandy, at least eleven villages in the duchy being so called.

The first part of the road to Bayeux passes through a flat and open district, resembling that on the other side of Caen; in the remaining half, the country is enclosed, with a more varied surface. Apple-trees again abound; and the old custom of suspending a bush over the door of an inn is commonly practised here. For this purpose misletoe is almost always selected. Throughout the whole of this district and the neighboring province of Brittany, the ancient attachment of the Druids to misletoe continues to a certain degree to prevail. The commencement of the new year is hailed by shouts of "au gui; l'an neuf;" and the gathering of the misletoe for the occasion is still the pretext for a merry-making, if not for a religious ceremony.

Bayeux was the seat of an academy of the Druids. Ausonius expressly addresses Attius Patera Pather, one of the professors at Bordeaux, as being of the family of the priesthood of this district:—

"Doctor potentum rhetorum,
Tu Bajocassis stirpe Druidarum satus;"

And tradition to this hour preserves the remembrance of the spot that was hallowed by the celebration of their mystic rites. This spot, an eminence adjoining the city, has subsequently served for the site of a priory dedicated to St. Nicholas de la chesnaye, thus commemorating by the epithet, the oaks that formed the holy grove. Near it stood the famous temple of Mount Phaunus, which was flourishing in the beginning of the fourth century, and, according to Rivet, was considered one of the three most celebrated in Gaul. Belenus was the divinity principally worshipped in it; but, according to popular superstition, adoration was also paid to a golden calf, which was buried in the hill, and still remains entombed there. Even within the last fifty years, two laborers have lost their lives in a fruitless attempt to find this hidden treasure. Tombs, and urns, and human bones, are constantly discovered; yet neither Druidic temples, nor pillars of stone, nor cromlechs or Celtic remains of any description exist, at least, at present, in the neighborhood of Bayeux.

Roman relics, however, abound. The vases and statues dug up near this city, have afforded employment to the pen and the pencil of Count Caylus, who, judging from the style of art, refers the greater part of them to the times of Julius and Augustus Cæsar. Medals of the earliest emperors have likewise frequently been detected among the foundations of the houses of the city; and even so recently as in the beginning of the present century, mutilated cippi, covered with Latin inscriptions, have been brought to light. These discoveries all tend to shew the Roman origin of Bayeux, and two Roman causeways also join here; so that, notwithstanding the arguments of the Abbé le Beuf, most antiquaries still believe that Bayeux was the city called by Ptolemy the Næomagus Viducassium.—The term Viducasses or Biducasses was in early ages changed to Bajocasses; and the city, following the custom that prevailed in Gaul, took the appellation of Bajocæ, or, as it was occasionally written, of Baiæ or Bagicæ. Its name in French has likewise been subject to alterations.—During the twelfth and thirteenth centuries, it was Baex and Bajeves; in the fourteenth Bajex; in the sixteenth Baieux; and soon afterwards it settled info the present orthography.

Pursuing the history of Bayeux somewhat farther, we find this city in the *Notitia Galileæ* holding the first rank among the towns of the *Secunda Lugdunensis*. During the Merovingian and Carlovingian dynasties, its importance is proved by the mint which was established here. Golden coins, struck under the first race of French sovereigns, inscribed *HBAJOCAS*, and silver pieces, coined by Charles the Bald, with the legend *HBAJOCAS-CIVITAS*, are mentioned by Le Blanc. Bayeux was also in those times, one of the head-quarters of the high functionaries, entitled *Missi Dominici*, who were annually deputed by the monarchy for the promulgation of their decrees and the administration of justice. Two other cities only in Neustria, Rouen and Lisieux, were distinguished with the same privilege.—Nor did Bayeux suffer any diminution of its honors, under the Norman Dukes: they regarded it as the second town of the duchy, and had a palace here, and frequently made it the seat of their *Aula Regio*.

The destruction of the Roman Bayeux is commonly ascribed, like that of the Roman Lisieux, to the Saxon invasion. No traces of the Viducassian capital are to be found in history, subsequently to the reign of Constantine; no medals, no inscriptions of a later period, have been dug up within its precincts. During the earliest incursions of the Saxons in Gaul, they seem to have made this immediate neighborhood the seat of a permanent settlement.

The Abbé Le Beuf places the district, known by the name of the Otlingua Saxonia, between Bayeux and Isigny; and Gregory of Tours, in his relation of the events that occurred towards the close of the sixth century, makes repeated mention of the Saxones Bajocassini, whom the early Norman historians style Saisnes de Bayeux. Under the reign of Charlemagne, a fresh establishment of Saxons took place here. That emperor, after the bloody defeat of this valiant people, about the year 804, caused ten thousand men, with their wives and children, to be delivered up to him as prisoners, and dispersed them in different parts of France. Some of the captives were colonized in Neustria; and, among the rest, Witikind, son of the brave chief of the same name, who had fought so nobly in defence of the liberty of his country, had lands assigned to him in the Bessin. Hence, names of Saxon origin commonly occur throughout the diocese of Bayeux; sometimes alone and undisguised, but more frequently in composition. Thus, in Estelan, you will have little difficulty in recognizing East-land: Cape la Hogue will readily suggest the idea of a lofty promontory; its appellation being derived from the German adjective, hoch, still written hoog, in Flemish: the Saxon word for the Almighty enters into the family names of Argot, Turgot, Bagot, Bigot, &c.; and, not to multiply examples, the quaking sands upon the sea-shore are to the present hour called bougues, an evident corruption of our own word bogs.

When, towards the middle of the same century, the Saxons were succeeded by the Normans, the country about Bayeux was one of the districts that suffered most from the new invaders. Two bishops of the see, Sulpitius and Baltfridus, were murdered by the barbarians; and Bayeux itself was pillaged and burned, notwithstanding the valiant resistance made by the governor, Berenger. This nobleman, who was count of the Bessin, was personally obnoxious to Rollo, for having refused him his daughter, the beautiful Poppea, in marriage. But, on the capture of the town, Poppea was taken prisoner, and compelled to share the conqueror's bed. Bayeux arose from its ruins under the auspices of Botho, a Norman chieftain, to whom Rollo was greatly attached, and who succeeded to the honors of Berenger. By him the town was rebuilt, and filled with a Norman population, the consequence of which was, according to Dudo of St. Quintin, that William Longa-Spatha, the successor of Rollo, who hated the French language, sent his son, Duke Richard, to be educated at Bayeux, where Danish alone was spoken. And the example of the Duke continued for some time to be imitated by his successors upon the throne; so that Bayeux became the academy for the children of the royal family, till they arrived at a sufficient age to be removed to the metropolis, there to be instructed in the art of government.

The dignity of Count of the Bessin ceased in the reign of William the Conqueror, in consequence of a rebellion on the part of the barons, which had well nigh cost that sovereign his life. From that time, till the conquest of Normandy by the French, the nobleman, who presided over the Bessin, bore the title of the king's viscount; and, under this name, you will find him the first cited among the four viscounts of Lower Normandy, in the famous parliament of all the barons of this part of the duchy, convened at Caen by Henry IInd, in 1152.—When Philip Augustus gained possession of Normandy, all similar appointments were re-modelled, and viscounts placed in every town; but their power was restricted to the mere administration of justice, the rest of their privileges being transferred to a new description of officers, who were then created, with the name of bailiffs. The bailiwicks assigned to these bore no reference to the ancient divisions of the duchy; but the territorial partition made at that time, has ever since been preserved, and Caen, which was honored by Philip with a preference over Bayeux, continues to the present day to retain the pre-eminence.

After these troubles, Bayeux enjoyed a temporary tranquillity; and, according to the celebrated historical tapestry and to the Roman de Rou, this city was selected for the place at which William the Conqueror, upon being nominated by Edward, as his successor to the crown of England, caused Harold to attend, and to do homage to him in the name of the nation. The oath was taken upon a missal covered with cloth of gold, in the presence of the prelates and grandees of the duchy; and the reliques of the saints were collected from all quarters to bear witness to the ceremony. Bayeux was also the spot in which Henry Ist was detained prisoner by his eldest brother, and it suffered for this unfortunate distinction; for Henry had scarcely ascended the English throne, when, upon a shallow pretext, he advanced against the city, laid siege to it, and burned it to the ground; whether moved to this act of vengeance from hatred towards the seat of his sufferings, or to satisfy the foreigners in his pay, whom the length of the siege had much irritated. He had promised these men the pillage of the city, and he kept his word; but the soldiers were not content with the plunder: they set fire to the town, and what had escaped their ravages, perished in the flames.[84] In 1356, under the reign of Edward IIIrd, Bayeux experienced nearly the same fate from our countrymen; and in the following century it again suffered severely from their arms, till the decisive battle of Formigny, fought within ten miles of the city, compelled Henry VIth to withdraw from Normandy, carrying with him scarcely any other trophies of his former conquests, than a great collection of Norman charters, and, among the rest, those of Bayeux, which are to this hour preserved in the tower of London.

During the subsequent wars occasioned by the reformation, this town bore its share in the common sufferings of the north of France. The horrors experienced by other places on the occasion were even surpassed by the outrages that were committed at Bayeux; but it is impossible to enter into details which are equally revolting to decency and to humanity.

Of late years, Bayeux has been altogether an open town. The old castle, the last relic of its military character, a spacious fortress flanked by ten square towers, was demolished in 1773; and, as the poet of Bayeux has sung[85],—

"... Gaulois, Romains, Saxons,

Oppresseurs, opprimés, colliers, faisceaux, blasons,

Tout dort. Du vieux château la taciturne enceinte

Expire. Par degrés j 'ai vu sa gloire éteinte.

J 'ai marché sur ses tours, erré dans ses fossés:

Tels qu 'un songe bientôt ils vont être effacés."

And in truth, they are so effectually *effaced*, that not a single vestige of the walls and towers can now be discovered.

Bayeux is situated in the midst of a fertile country, particularly rich in pasturage. The Aure, which washes its walls, is a small and insignificant streamlet, and though the city is within five miles of the sea, yet the river is quite useless for the purposes of commerce, as not a vessel can float in it. The present population of the town consists of about ten thousand inhabitants, and these have little other employment than lace-making.—Bayeux wears the appearance of decay: most of the houses are ordinary; and, though some of them are built of stone, by far the greater part are only of wood and plaster. In the midst, however, of these, rises the noble cathedral; but this I shall reserve for the subject of my next letter, concluding the present with a few remarks upon that matchless relic, which,

"... des siècles respecté,

En peignant des héros honore la beauté."

The very curious piece of historical needle-work, now generally known by the name of the Bayeux tapestry, was first brought into public notice in the early part of the last century, by Father Montfaucon and M. Lancelot, both of whom, in their respective publications, the Monumens de la Monarchie Française[86], and a paper inserted in the Mémoires de l'Académie des Inscriptions[87], have figured and described this celebrated specimen of ancient art. Montfaucon's plates were afterwards republished by Ducarel[88], with the addition of a short dissertation and explanation, by an able antiquary of our own country, Smart Lethieuilier.

These plates, however, in the original, and still more in the copies, were miserably incorrect, and calculated not to inform, but to mislead the inquirer. When therefore the late war was concluded and France became again accessible to an Englishman, our Society of Antiquaries, justly considering the tapestry as being at least equally connected with English as with French history, and regarding it as a matter of national importance, that so curious a document should be made known by the most faithful representation, employed an artist, fitted above all others for the purpose, by his knowledge of history and his abilities as a draughtsman, to prepare an exact fac-simile of the whole. Under the auspices of the Society, Mr. C.A. Stothard undertook the task; and he has executed it in the course of two successive visits with the greatest accuracy and skill. The engravings from his drawings we may hope shortly to see: meanwhile, to give you some idea of the original, I enclose a sketch, which has no other merit than that of being a faithful transcript. It is reduced one half from a tracing made from the tapestry itself. By referring to Montfaucon, you will find the figure it represents under the fifty-ninth inscription in the original, where "a knight, with a private banner, issues to mount a led horse." His beardless countenance denotes him a Norman; and the mail covering to his legs equally proves him to be one of the most distinguished characters.

Within the few last years this tapestry has been the subject of three interesting papers, read before the Society of Antiquaries. The first and most important, from the pen of the Abbé de la Rue[89], has for its object the refutation of the opinions of Montfaucon and Lancelot, who, following the commonly received tradition, refer the tapestry to the time of the conquest, and represent it as the work of Queen Matilda and her attendant damsels. The Abbé's principal arguments are derived from the silence of contemporary authors, and especially of Wace, who was himself a canon of Bayeux;—from its being unnoticed in any charters or deeds of gift connected with the cathedral;—from the improbability that so large a roll of such perishable materials would have escaped destruction when the cathedral was burned in 1106;—from the unfinished state of the story;—from its containing some Saxon names unknown to the Normans;—and from representations taken from the fables of Æsop being worked on the borders, whereas the northern parts of Europe were not made acquainted with these fables, till the translation of a portion of them by Henry Ist, who thence obtained his surname of Beauclerk.—These and other considerations, have led the learned Abbé to coincide in opinion with Lord Littleton and Mr. Hume, that the tapestry is the production of the Empress Maud, and that it was in reality wrought by natives of our own island, whose inhabitants were at that time so famous for labors of this description, that the common mode of expressing a piece of embroidery, was by calling it an English work.

The Abbé shortly afterwards found an opponent in another member of the society, Mr. Hudson Gurney, who, without following his predecessor through the line of his arguments, contented himself with briefly stating the three following reasons for ascribing the tapestry to Matilda, wife to the Conqueror[90].—First, that in the many buildings therein pourtrayed, there is not the least appearance of a pointed arch, though much pointed work is found in the ornaments of the running border; whilst, on the contrary, the features of Norman architecture, the square buttress, flat to the walls, and the square tower surmounted by, or rather ending in, a low pinnacle, are therein frequently repeated.—Secondly, that all the knights are in ring armour, many of their shields charged with a species of cross and five dots, and some with dragons, but none with any thing of the nature of armorial bearings, which, in a lower age, there would have been; and that all wear a triangular sort of conical helmet, with a nasal, when represented armed.—And, Thirdly, that the Norman banner is, invariably, Argent, a Cross, Or, in a Bordure Azure; and that this is repeated over and over again, as it is in the war against Conan, as well as at Pevensey and at Hastings; but there is neither hint nor trace of the later invention of the Norman leopards.—Mr. Gurney's arguments are ingenious, but they are not, I fear, likely to be considered conclusive: he however, has been particularly successful in another observation, that all writers, who had previously treated of the Bayeux tapestry, had called it a Monument of the Conquest of England; following, therein, M. Lancelot, and speaking of it as an unfinished work, whereas, it is in fact an apologetical history of the claims of William to the crown of England, and of the breach of faith and fall of Harold, in a perfect and finished action.—With this explanation before us, aided by the short indication that is given of the subjects of the seventy-two compartments of the tapestry, a new light is thrown upon the story.

The third memoir is from the pen of Mr. Amyot, and concludes with an able metrical translation from Wace. It is confined almost exclusively to the discussion of the single historical fact, how far Harold was really sent by the Confessor to offer the succession to William; but this point, however interesting, in itself, is unconnected with my present object: it is sufficient for me to shew you the various sources from which you may derive information upon the subject.

Supposing the Bayeux tapestry to be really from the hands of the Queen, or the Empress, (and that it was so appears to me proved by internal evidence,) it is rather extraordinary that the earliest notice which is to be found of a piece of workmanship, so interesting from its author and its subjects, should be contained in an inventory of the precious effects deposited in the treasury of the church, dated 1476. It is also remarkable that this inventory, in mentioning such an article, should call it simply a very long piece of cloth, embroidered with figures and writing, representing the conquest of England, without any reference to the royal artist or the donor.

Observations of this nature will suggest themselves to every one, and the arguments urged by the Abbé de la Rue are very strong; and yet I confess that my own feelings always inclined to the side of those who assign the highest antiquity to the tapestry. I think so the more since I have seen it. No one appears so likely to have undertaken such a task as the female most nearly connected with the principal personage concerned in it, and especially if we consider what the character of this female was: the details which it contains are so minute, that they could scarcely have been known, except at the time when they took place: the letters agree in form with those upon Matilda's tomb; and the manners and customs of the age are also preserved.—Mr. Stothard, who is of the same opinion as to the date of the tapestry, very justly observes, that the last of these circumstances can scarcely be sufficiently insisted upon; for that "it was the invariable practice with artists in every country, excepting Italy, during the middle ages, whatever subject they took in hand, to represent it according to the costume of their own times."

Till the revolution, the tapestry was always kept in the cathedral, in a chapel on the south side, dedicated to Thomas à Becket, and was only exposed to public view once a year, during the octave of the feast of St. John on which occasion it was hung up in the nave of the church, which it completely surrounded. From the time thus selected for the display of it, the tapestry acquired the name of le toile de Saint Jean; and it is to the present day commonly so called in the city. During the most stormy part of the revolution, it was secreted; but it was brought to Paris when the fury of vandalism had subsided. And, when the first Consul was preparing for the invasion of England, this ancient trophy of the subjugation of the British nation was proudly exhibited to the gaze of the Parisians, who saw another Conqueror in Napoléon Bonaparté; and many well-sounding effusions, in prose and verse, appeared, in which the laurels of Duke William were transferred, by anticipation, to the brows of the child and champion of jacobinism. After this display, Bonaparté returned the tapestry to the municipality, accompanied by a letter, in which he thanked them for the care they had taken of so precious a relic. From that period to the present, it has remained in the residence appropriated to the mayor, the former episcopal palace; and here we saw it.

It is a piece of brownish linen cloth, about two hundred and twelve feet long, and eighteen inches wide, French measure. The figures are worked with worsted of different colors, but principally light red, blue, and yellow. The historical series is included between borders composed of animals, &c. The colors are faded, but not so much so as might have been expected. The figures exhibit a regular line of events, commencing with Edward the Confessor seated upon his throne, in the act of dispatching Harold to the court of the Norman Duke, and continued through Harold's journey, his capture by the Comte de Ponthieu, his interview with William, the death of Edward, the usurpation of the British throne by Harold, the Norman invasion, the battle of Hastings, and Harold's death. These various events are distributed into seventy-two compartments, each of them designated by an inscription in Latin. Ducarel justly compares the style of the execution to that of a girl's sampler. The figures are covered with work, except on their faces, which are merely in outline. In point of drawing, they are superior to the contemporary sculpture at St. Georges and elsewhere; and the performance is not deficient in energy. The colors are distributed rather fancifully: thus the fore and off legs of the horses are varied. It is hardly necessary to observe that perspective is wholly disregarded, and that no attempt is made to express light and shadow.

Great attention, however, is paid to costume; and more individuality of character has been preserved than could have been expected, considering the rude style of the workmanship. The Saxons are represented with long mustachios: the Normans have their upper lip shaven, and retain little more hair upon their heads than a single lock in front. Historians relate how the English spies reported the invading army to be wholly composed of ecclesiastics; and this tapestry affords a graphical illustration of the chroniclers' text. Not the least remarkable feature of the tapestry, in point of costume, lies in the armor, which, in some instances, is formed of interlaced rings; in others, of square compartments; and in others, of lozenges. Those who contend for the antiquity of Duke William's equestrian statue at Caen, may find a confirmation of their opinions in the shape of the saddles assigned to the figures of the Bayeux tapestry; and equally so in their cloaks, and their pendant braided tresses.

The tapestry is coiled round a cylinder, which is turned by a winch and wheel; and it is rolled and unrolled with so little attention, that if it continues under such management as the present, it will be wholly ruined in the course of half a century. It is injured at the beginning: towards the end it becomes very ragged, and several of the figures have completely disappeared. The worsted is unravelling too in many of the intermediate portions. As yet, however, it is still in good preservation, considering its great age, though, as I have just observed, it will not long continue so. The bishop and chapter have lately applied to government, requesting that the tapestry may be restored to the church. I hope their application will be successful.

Footnotes:

[82] The most interesting relic of Roman times yet found at Vieux, is a cippus of variegated marble, about five feet high by two feet wide, and bearing inscriptions upon three of its sides. It generally passes in France by the name of the Torigny marble, being preserved at the small town of the latter name, whither it was carried in 1580, the very year when it was dug up. The Abbé Le Beuf has made it the subject of a distinct paper in the Mémoires de l'Académie des Inscriptions. This cippus supported a statue raised in honor of Titus Sennius Sollemnis, a Viducassian by birth, and one of the high priests of the town. The statue was erected to him after his death, in the Viducassian capital, upon a piece of ground granted by the senate for the purpose, in pursuance of a general decree passed by the province of Gaul. The inscriptions set forth the motives that induced the nation to bestow so marked a

distinction upon a simple individual; and, in the foremost rank of his merits, they place the games which he had given to his fellow-citizens, during four successive days.

[83] Mémoires de l'Académie des Inscriptions, XXI. p. 489.

[84] Archæologia, XVII. p. 911.

[85] Bayeux et ses Environs, par M. Delauney, p. 12.

[86] I. p. 371-379; pl. 35-49, and II. p. 1-29; pl. 1-9.

[87] VI. p. 739, and VIII. p. 602.

[88] Anglo-Norman Antiquities, Appendix, No. 1.

[89] Archæologia, XVII. p. 85.

[90] Archæologia, XVIII. p. 359.

Sculptures at Bayeux

LETTER XXVIII
CATHEDRAL OF BAYEUX—CANON OF CAMBREMER—COPE OF ST. REGNOBERT—ODO

(*Bayeux, August,* 1818.)

Excepting the tapestry and the cathedral, Bayeux, at this time, offers no objects of interest to the curious traveller. Its convents are either demolished, or so dilapidated or altered, that they have lost their characteristic features; and its eighteen parish churches are now reduced to four. We wandered awhile about the town, vainly looking after some relic of ancient art, to send you by way of a memento of Bayeux. At length, two presented themselves—the entrance of the corn-market, formerly the chapel of St. Margaret, a Norman arch, remarkable for the lamb and banner, an emblem of the saint, sculptured on the transom stone; and a small stone tablet, attached to an old house near the cathedral. The whimsical singularity of the latter, induced us to give it the preference. It may possibly be of the workmanship of the fourteenth century, and possibly much later. In all probability, it owes its existence merely to a caprice on the part of the owner of the residence, whose crest may be indicated by the tortoises which surmount the columns by way of capitals. Still there is merit in the performance, though perhaps for nothing so much as for the accurate resemblance of peeled wood; and this I never saw imitated with equal fidelity in stone.

But, however unattractive Bayeux may be in other respects, so long as the cathedral is suffered to stand, the city will never want interest. It is supposed that the first church erected here was built by St. Exuperius otherwise called St. Suspirius, or St. Spirius, who, according to the distich subjoined to his portrait, formerly painted on one of the windows of the nave, was not only the earliest bishop of the diocese, but claimed the merit of having introduced the Christian faith into Normandy,—

"Primitùs hic pastor templi fuit hujus et auctor,

Catholicamque fidem Normannis attulit idem."

St. Exuperius lived in the third century, and his efforts towards the propagation of the gospel were attended with so great success, that his successor, St. Regnobert, was obliged to take down the edifice thus recently raised, and to re-construct it on a more enlarged scale, for the purpose of accommodating the increasing congregation. Regnobert is likewise reported to have built the celebrated chapel on the sea-coast, dedicated to our Lady de la Délivrande; and the people believe that a portion at least, of both the one and the other of these original edifices, exists to the present day. The Abbé Béziers, however, in his *History of Bayeux*, maintains, and with truth, that St. Regnobert's cathedral was destroyed by the Normans; and he adds that, immediately after the conversion of Rollo, another was raised in its stead on the same spot, and that this latter was one of those which the chieftain most enriched by his endowments at the period of his baptism.

A dreadful fire, in the year 1046, reduced the Norman cathedral to ashes; but the episcopal throne was then filled by a prelate who wanted neither disposition nor abilities to repair the damage. Hugh, the third bishop of that name, son to Ralph, Count of the Bessin, who, by the mother's side, was brother to Duke Richard Ist, presided at that time over the see of Bayeux. Jealous for the honor of his diocese, the prelate instantly applied himself to rebuild the cathedral; but he lived to see only a small progress made in his work. It was finished by a prelate of still greater, though evil celebrity, the unruly Odo, brother to the Conqueror, who, for more than fifty years, continued bishop of this see, and by his unbounded liberality and munificence in the discharge of his high office, proved himself worthy of his princely descent. The Conqueror and his queen, attended by their sons, Robert and William, and by the archbishops of Canterbury and York, as well as by the various bishops and barons of the province, were present at the dedication of the church, which was performed in 1077, by John, Archbishop of Rouen. Odo, on the occasion, enriched his church with various gifts, one of which has been particularly recorded. It was a crown of wood and copper, sixteen feet high and thirty-eight feet in diameter, covered with silver plates, and diversified with other crowns in the shape of towers; the whole made to support an immense number of tapers, that were lighted on high festivals. This crown was suspended in the nave, opposite the great crucifix; and it continued to hang there till it was destroyed by the Huguenots, in 1562.

It is doubtful how much, or indeed if any portion, of the church erected by Odo be now in existence. Thirty years had scarcely elapsed from the date of its dedication, when, as I have already mentioned to you, the troops of Henry Ist destroyed Bayeux with fire. The ruin was so complete, that for more than fifty years, no attempt was made to re-construct the cathedral; but it remained in ashes until the year 1157, when bishop, Philip of Harcourt, determined to restore it. A question has arisen whether the oldest part of what is now standing, be the work of Philip or of Odo. The lapse of eighty years in those early times, would perhaps occasion no very sensible difference in style, and chroniclers do not afford the means of determining, if, at the time when Bayeux suffered so dreadfully in 1106, the church was actually burned to the ground, or only materially damaged. In the History of the Diocese we are merely told that Philip, having, by means of papal bulls, happily succeeded in regaining possession of all the privileges, honors, and property of the see, began to rebuild his cathedral in 1159, and completed it with great glory and expence.—From that time forward, we hear no more of demolition or of re-edification; but the injuries done by the silent lapse of ages, and the continued desire on the part of the prelates to beautify and to enlarge their church, have produced nearly the same effect as fire or warfare. The building, as it now stands, is a medley of various ages; and, in the absence of historical record, it would be extremely difficult to define the several portions that are to be assigned to each.

The west front is flanked by two Norman towers, bold and massy, with semi-circular arches in the highest stories. The spires likewise appear ancient, though these and the surrounding pinnacles are all gothic. The northern one, according to tradition, was built with the church; the southern, in 1424. They both greatly resemble those of the abbey-church of St. Stephen at Caen. But the whole centre of this front, and indeed both the sides also, as high as the roof, is faced by a screen divided into five compartments. In the middle is a large, wide, pointed arch, with a square-headed entrance beneath. North and south of this are deep arches, evidently older, but likewise pointed, having their sides above the pillars, and the flat arched part of the door-way, filled with small figures. The door-ways themselves are arches that occupy only one half of the width of those which enclose them. In the two exterior compartments the arches are unpierced, and are flanked by a profusion of clustered pillars. Over each of the four lateral arches, rises a crocketed pyramid: the central one is surmounted by a flat balustrade, above which, behind the screen, is a large pointed window, and over it a row of saints, standing under trefoil-headed arches, arranged in pairs, the pediment terminating above each pair of arches in a pyramidal canopy.

The outside of the nave is of florid gothic, but it is not of a pure style; nor is the southern portal, which, nevertheless, considered as a whole, is bold and appropriate. On each side of the door-way were originally three statues, whose tabernacles remain, though the saints have been torn out of the niches. Over the door is a bas-relief, containing numerous figures disposed in three compartments, and representing some legendary tale, which our knowledge of that kind of lore would not enable us to decipher.—

The exterior of the choir is likewise of pointed architecture: it is considerably more simple, and excels, in this respect, the rest of the church. But even here there is a great want of uniformity: some of the windows are deeply imbedded in the walls; others are nearly on a level with their surface.—The cupola, which caps the low central tower, is wretchedly at variance with the other parts of the building. It was erected in the year 1714, at the expence of the bishop, Francis de Nesmond; and it is, as might be expected from a performance of that period, rather Grecian than gothic. Whichever style it may be termed, it is a bad specimen of either. And yet, such as it is, we are assured by Béziers, that it was built after the designs of a celebrated architect of the name of Moussard, and that it excited particular attention, and called forth loud praises, on the part of the Maréchal de Vauban, who was, probably, a better judge of a modern fortification, than of a gothic cathedral.

The interior of the church consists of a wide nave, with side-aisles, and chapels beyond them. The first six piers of the nave are very massy, and faced with semi-circular pillars supporting an entablature. The arches above them are Norman, encircled with rich bands, composed chiefly of the chevron moulding and diamonds. On one of them is a curious border of heads, as upon the celebrated door-way at Oxford; but the heads at Bayeux are of much more regular workmanship and more distinctly defined. Had circumstances allowed, I would have sent you an accurate drawing of them; but our time did not permit such a one to be made, and I must beg of you to be contented with the annexed slight sketch.

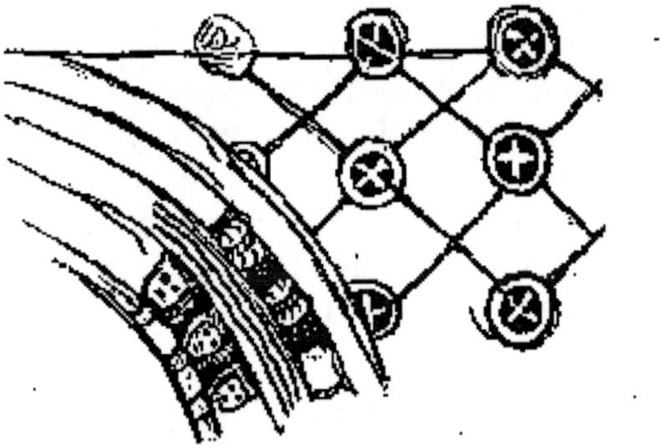

The wall above the arches is incrusted with a species of tessellated work of free-stone, of varied patterns, some interwoven, others reticulated, as seen in the sketches: the lines indented in the stones, as well as the joints which form the patterns, are filled with a black cement or mastich, so as to form a kind of *niello.*

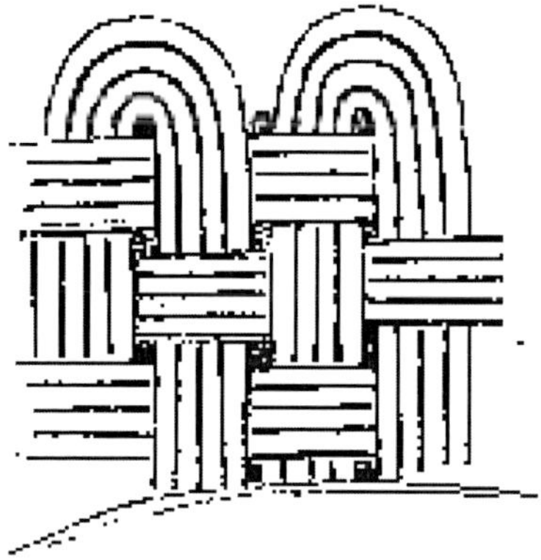

With the sixth arch of the nave begins the pointed style. The capitals of the pillars are complicated, and the carving upon them is an evident attempt at an imitation of the Grecian orders. In this part of the church there is no triforium; but a row of small quartrefoils runs immediately above the ornaments of the spandrils; and above the quatrefoils is a cornice of an antique pattern, which is surmounted by a light gallery in front of the windows of the clerestory, the largest windows I remember to have seen in a similar situation. They extend almost from the roof to the line of the old Norman basement. Their magnitude is rendered still more remarkable by their being arranged in pairs, each separate pair inclosed within a pointed arch, and its windows parted only by a clustered pillar. The very lofty arches that support the central tower, are likewise pointed; as are those of the transepts, the choir, the side-aisles, and the chapels. In short, excepting the arches immediately beneath the northern and southern towers, which are most probably relics of Odo's cathedral, the part of the nave, which I first described, is all that is left above-ground of the semi-circular style; and this is of a very different character from whatever else I have seen of Norman architecture. The circular ornaments inserted in the spandrils of the arches of the choir, possess, as a friend of mine observes, somewhat of the Moorish, or, perhaps, Tartarian character; being nearly in the style of the ornaments which are found in the same situation in the Mogul mosques and tombs, though here they have much more flow and harmony in the curves. Some are merely in bas-relief: in others the central circles are deeply perforated, whilst the ribs are composed of delicate tracery.—There are so many peculiarities both in the arrangement and in the details of this cathedral[91], that it is quite impossible to convey an adequate idea of them by a verbal description; and I can only hope that they will be hereafter made familiar to the English antiquarian by the pencil of Mr. Cotman or Mr. Stothard.

The screen that separates the nave from the choir is Grecian, and is as much at variance with the inside of such a church, as the cupola, which is nearly over it, is with the exterior.—Upon the roof of the choir, are still to be seen the portraits of the first twenty-one bishops of Bayeux, each with his name inscribed by his side. The execution of the portraits is very rude, particularly that of the twelve earliest, whose busts are represented. The artist has contented himself with exhibiting the heads only, of the remaining nine. Common tradition refers the whole of these portraits to the time of Odo; but it is hardly necessary to observe, that the groined and pointed vaulting is subsequent to his date.—Bayeux cathedral abounded in works of this description of art: the walls of the chapels of the choir were covered with large fresco-paintings, now nearly obliterated.—It is believed, and with every appearance of probability, that the Lady-Chapel was erected at a time posterior to the rest of the building; but there is no certain account of its date. Before the revolution, it served as a burial-place for some of the bishops of the see, and for a duke of the noble family of Montemart. Their tombs ornamented the chapel, which now appears desolate and naked, retaining no other of its original decorations, than a series of small paintings, which represent the life of the Holy Virgin, and are deserving of some attention from the character of expression in the faces, though the drawing in general is bad. Over the altar is a picture, in which an angel is pointing out our Savior and the Virgin to a dying man, whose countenance is admirable.—The stalls of the choir display a profusion of beautiful oak carving; and beneath them are sculptured misereres, the first which we have observed in Normandy.—Very little painted glass is to be found in any part of the church; but the glazing of the windows is composed of complicated patterns. This species of ornament was introduced about the time of Louis XIVth; and Felibien, who has given several pattern plates in his treatise on architecture, observes, that it was intended to supply the place of painted glass, which, as it was then thought, excluded the light.

Beneath the choir is a subterraneous chapel dedicated to St. Maimertus, otherwise called St. Manvieu. Its character is so similar to that of the crypt at the abbey of the Holy Trinity at Caen, that there would be little risk in pronouncing it to be part of Odo's church. It is supported on twelve pillars, disposed in two rows, the last pillar of each row being imbedded in the wall. The capitals of the pillars are carved, each with a different design from the rest. Their sculpture bears a strong resemblance to some of what is seen in similar situations in the Egyptian temples; indeed, so strong, that a very able judge tells me he has been led to suspect that the model might have been introduced by an anchorite from the desert. Take the following as a specimen.

The walls of the crypt are covered with paintings, probably of the fifteenth century; but those upon the springing of the arches above the pillars, appear considerably older. Each spandril contains an angel, holding a trumpet or other musical instrument. The outlines of these figures are strongly drawn in black.—Upon the right-hand side, on entering the chapel, is the altar-tomb of John de Boissy, who was bishop at the beginning of the fifteenth century; and, on the opposite side, stands that of his immediate predecessor, Nicolas de Bosc. Their monuments were originally ornamented with bas-reliefs and paintings, all which were mutilated and effaced during the religious wars. De Boissy's effigy, however, remains, though greatly injured; and the following epitaph to his memory is preserved in a perfect state, over the only window that gives light to this crypt. The inscription is curious, as recording the discovery of the chapel, which had been forgotten and unknown for centuries.

"En l'an mil quatrc cens et douze
Tiers jour d 'Avril que pluye arrouse
Les biens de la terre, la journée
Que la Pasques fut célébrée
Noble homme et révérend père
Jehan de Boissy, de la mère
Eglise de Bayeux Pasteur
Rendi l 'âme à Son Créateur
Et lors en foillant la place
Devant le grant autel de grâce
Trova l 'on la basse chapelle
Dont il n 'avoit esté nouvelle
Ou il est mis en sépulture
Dieu veuille avoir son âme en cure,—Amen."

This inscription is engraved as prose: verse is very frequently written in this manner in ancient manuscripts, which custom, as Joseph Ritson conjectured, arose "from a desire of promoting the salvation of parchment." I must also add, that the initial letters are colored red and blue, so that the whole bears a near resemblance to a manuscript page.

There is another epitaph, engraved in large letters, upon the exterior of the southern tower, which is an odd specimen of the spirit of the middle ages. It is supposed to have been placed there in the twelfth century.

"Quarta dies Pasche fuerat cum Clerus ad hujus
Que jacet hic vetule venimus exequias:
Letitieque diem magis amisisse dolemus
Quam centum tales si caderent vetule."

Some authors contend, that the old lady alluded to was the mistress of one of the Dukes of Normandy: others believe her to have been the *chère amie* of Robert, Earl of Gloucester, illegitimate son to Henry Ist.

Till lately, there was an epitaph within the church, which, without containing in itself any thing remarkable, strange, or mysterious, had a legend connected: with it, that supplied the verger with an inexhaustible fund of entertainment for the curious and the credulous. The epitaph simply commemorated John Patye, canon of the prebend of Cambremer, who died in 1540; but upon the same plate of copper with the inscription, was also engraved the Virgin, with John Patye at her feet, kneeling, and apparently in the act of reading from a book placed on a fald-stool. Behind the priest stood St. John the Baptist, the patron saint of the prebend, having one hand upon his votary 's neck, while with the other he pointed to a lamb.—In all this, there was still nothing remarkable: unfortunately, however, the artist, wishing perhaps to add importance to the saint, had represented him of gigantic stature; and hence originated the story, which continues to the present day, to frighten the old women, and to amuse the children of Bayeux.—

> Once upon a time, the wicked canons of the cathedral murdered their bishop; in consequence of which foul deed, they and their successors for ever, were enjoined, by way of penance, annually to send one of their number to Rome, there to chaunt the epistle at the midnight mass. In the course of revolving centuries, this vexatious duty fell to the turn of the canon of Cambremer, who, to the surprise of the community, testified neither anxiety nor haste on the occasion.—Christmas-eve arrived, and the canon was still in his cell: Christmas-night came, and still he did not stir. At length, when the mass was actually begun, his brethren, more uneasy than himself, reproached him with his delay; upon which he muttered his spell, called up a spirit, mounted him, reached Rome in the twinkling of an eye, performed his task, and, the service being ended, he stormed the archives of the Vatican, where he burned the compulsory act, and then returned by the same conveyance to Bayeux, which he reached before the mass was completed, and, to the unspeakable joy of the chapter, announced the happy tidings of their deliverance.

So idle and unmeaning is the tale, that I should scarcely have thought it worth while to have repeated it, but for the Latin distich, which, as the story goes, was extemporized by the demon, at the moment when they were flying over the Tuscan sea, and by which he sought to mislead his rider, and to cause him to end his journey beneath the deep.—The sense of the verses is not very perspicuous, but they are remarkable for reading forwards and backwards the same; and though to you they may appear a childish waste of intellect, you will, I am sure, admit them to be ingenious, and they may amuse some of the younger members of your family:—

"Signa te, signa, temerè me tangis et angis;

Roma tibi subito motibus ibit amor."—

I must dismiss the canon of Cambremer, by stating, that I am informed by a friend, that the same story is also found in the lives of sundry other wizards and sorcerers of the good old times.

Bayeux cathedral, like the other Neustrian churches, has been deprived of its sainted relics, and its most precious treasures, in consequence of the successive spoliations which have been inflicted upon it by heathen Normans, heretical Calvinists, and philosophical jacobins. The body of St. Exuperius was carried, in the ninth century, for safety to Corbeil, and the chapter have never been able to recover it: that of St. Regnobert was in after times stolen by the Huguenots. Many are the attempts that have been made to regain the relics of the first bishop of the see; but the town of Corbeil retained possession, whilst the Bajocessians attempted to console themselves by antithetical piety.—"Referamus Deo gratias, nec inde aliquid nos minus habere credamus, quòd Corbeliensis civitas pignus sacri corporis vindicavit. Teneant illi tabernaculum beatæ animæ in cineribus suis; nos ipsam teneamus animam in virtutibus suis: teneant illi ossa, nos merita: apud illos videatur remansisse quod terræ est, nos studeamus habere quod coeli est: amplectantur illi quod sepulchre, nos quod Paradiso continetur. Meminerit et beatior ille vir, utrique quidem loco, sed huic speciali se jure deberi."—St. Regnobert's chasuble is however, left to the church, together with his maniple and his stole, all of them articles of costly and elaborate workmanship. They were found in his coffin, when it was opened by the Calvinists; and they are now worn by the bishop, on the anniversary of the saint, as well as on five other high festivals, during the year; at which times, the faithful press with great devotion to kiss them. When not in use, they are kept in an ivory chest, magnificently embossed with solid silver, and bearing an inscription in the Cufic character, purporting that whatever honor men may have given to God, they cannot honor him so much as He deserves. Father Tournemine, the Jesuit, is of opinion, that this box was taken by the French troops, under Charles Martel, in their pillage of the Saracen camp, at the time of the memorable defeat of the infidels; and that it was afterwards presented to Charles the Bald, whose queen, Hermentrude, devoted it to the pious purpose of holding the relics of Regnobert, in gratitude for a cure which the monarch had received through the intercession of the saint. But this is merely a conjecture, and it is not improbable but that the chest may have been brought from Sicily, which abounded with Arabic artificers, at the time when it was occupied by the Normans.

St. Regnobert, who was one of the most illustrious bishops of Bayeux, is placed second on the list, in the History of the Diocese; but in the Gallia Christiana he stands twelfth in order. It was customary before the revolution, and it possibly may be so at present, for the inhabitants of the city, upon the twenty-fourth of October, the anniversary of his feast, to bring their domestic animals in solemn procession to the church, there to receive the episcopal benediction, in the same manner as is practised by the Romans with their horses, on the feast of St. Anthony.—St. Lupus, the fourth bishop, and St. Lascivus, the tenth, are remarkable for their names. St. Lupus is said to have been so called from his having destroyed the wolves in the vicinity of Bayeux[92]; and the other is reported to have been descended from the same person, whom Ausonius addresses in the following stanza, which has likewise been applied to this bishop.

"Iste *Lascivus* patiens vocari,
Nomen indignum probitate vitae
Abnuit nunquam; quia gratum ad aures
Esset amicas."—

But neither among her ancient nor her modern prelates can Bayeux boast of a name equally distinguished as that of Odo. Many were unquestionably the misdeeds of this great man, and many were probably his crimes, but no one who wore the episcopal mitre, ever deserved better of the see. As a statesman, Odo bore a leading part in all the principal transactions of the times: as a soldier, he accompanied the Conqueror to England, fought by his side at Hastings, and by his eloquence and his valor, contributed greatly to the success of that memorable day. Nor was William tardy in acknowledging the merits of his brother; for no sooner did he find himself seated firmly on the throne, than he rewarded Odo with the earldom of Kent, and appointed him his viceroy in England, whilst he himself crossed the channel, to superintend his affairs in Normandy. But the mind which was proof against difficulties, yielded, as too commonly happens, to prosperity. Nothing less than the papacy could satisfy the ambition of Odo: he abused the power with which he was invested in a flagrant manner; and William, finally, disgusted with his proceedings, arrested him with his own hand, and committed him prisoner to the old palace at Rouen, where he continued till the death of the monarch.—The sequel of the story is of the same complexion: more plots, attended now with success, and now with disgrace; till at length the prelate resolved to expiate his sins by a pilgrimage to the Holy Land, and died on his journey, at Palermo.—Such was Odo in his secular character: as a churchman, historians unanimously agree that he was most zealous for the honor of his diocese, indefatigable in re-building the churches which time or war had destroyed, liberal in endowments, munificent in presents, and ever anxiously intent upon procuring a supply of able ministers, establishing regular discipline, and reforming the morals of the flock committed to his charge.

The Bishop of Bayeux has at all times claimed the distinction of being regarded the first among the suffragan bishops of the Norman church. In the absence of the archbishop, he presides at, the ecclesiastical assemblies and councils. His revenue, before the revolution, was estimated at one hundred thousand livres: per annum. The see, in point of antiquity, even contests for the priority with Rouen. From time immemorial, the chapter has enjoyed the right of mintage; and they appear to have used it till the year 1577, at which time their coin was so much counterfeited, that they were induced to recal it by public proclamation. Their money, which was of the size of a piece of two sous, was stamped, on one side, with a two-headed eagle, and the legend moneta capituli; and on the obverse, with the letter V, surrounded by the word Bajocensis. The eagle was probably adopted, in allusion to the arms of the see, which were, gules; an eagle displayed with two heads, or[93].—Another privilege of the chapter was, that no person of illegitimate birth could be allowed to hold place in it, under any pretext or dispensation whatever.—Among their peculiar customs, they imitated that of the see of Rouen, in the annual election of a boy-bishop upon Innocents'-day; a practice prevalent in many churches in Spain and Germany, and notoriously in England at Salisbury. The young chorister took the crozier in his hands, during the first vespers, at the verse in the Magnificat, "He has put down the mighty from their seats, and has exalted the humble and meek;" and he resigned his dignity at the same verse in the second vespers.—The ceremony was abolished in 1482.

Footnotes:

[91] The following are the dimensions of the church, in French measure, according to Béziers.

	FEET.
Height of the central tower	**224**
Ditto of the two western ditto	**230**
Length of the interior of the church	**296**
Width of ditto	**76**

Height of ditto	**76**
Length of the nave	**140**
Width of ditto	**38**
Ditto of side-aisles	**17**
Ditto of chapels	**15**
Length of the transepts	**113**
Width of ditto	**33**
Length of the choir	**118**
Width of ditto	**36**

[92] A new St. Lupus is now wanted for the see; for wolves are by no means extinct in the neighborhood of Bayeux. We saw a tame one, kept near the cathedral, which had been taken in the woods, about a year ago, when it was quite young. Wild boars are likewise found in considerable numbers, and the breed is encouraged for the purposes of hunting.

[93] In its origin, the Baiocco of Naples seems to have been the two-penny piece of Bayeux, its denomination being abbreviated from the last word in the legend. It has been supposed that the coin was struck and named by lusty Joan, as a token of her affection towards a Frisick warrier, who, in his own country, was called the Boynke, or the Squire; but we think that our etymology is the most natural one.

LETTER XXIX
CHURCH AND CASTLE OF CREULLY—FALAISE—CASTLE—CHURCHES-FAIR OF GUIBRAY

(*Falaise, August,* 1818.)

Previously to quitting Bayeux, we paid our respects to M. Pluquet, a diligent antiquary, who has been for some time past engaged in writing a history of the city. His collections for this purpose are extensive, and the number of curious books which he possesses is very considerable. Amongst those which he shewed to us, the works relating to Normandy constituted an important portion. His manuscript missals are numerous and valuable. I was also much pleased by the inspection of an old copy of Aristophanes, which had formerly belonged to Rabelais, and bore upon its title-page the mark of his ownership, in the hand-writing of the witty, though profligate, satirist himself. M. Pluquet's kindness allowed me to make the tracing of the signature, which I send you.—

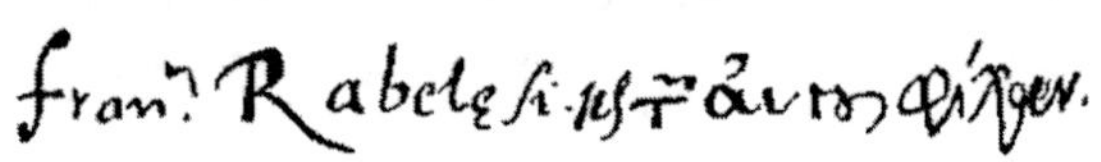

Such an addition as we here find to Rabelais' name, denoting that the owner of a book considered it as being the property of his friends conjointly with himself, is not of uncommon occurrence. Our friend, Mr. Dibdin, who had been here shortly before us, and had carried off, as we were told, some works of great rarity from this collection, has enumerated more than one instance of the kind in his Bibliographical Decameron; and the valuable library of my excellent friend, Mr. Sparrow, of Worlingham, contains an Erasmus, which was the property of Sir Thomas Wotton, and bears, stamped upon its covers, Thomae Wotton et amicorum.

From Bayeux we returned to Caen, by way of Creully, passing along bad roads, through an open, uninteresting country, almost wholly cropped with buck-wheat.—The barony of Creully was erected by Henry Ist, in favor of his natural son, the Earl of Gloucester: it was afterwards held by different noble families, and continued to be so till the time of the revolution. At that period, it gave a title to a branch of the line of Montmorenci, whose emigration caused the domain to be confiscated, and sold as national property; but the baronial castle is still standing, and displays, in two of its towers and in a chimney of unusual form, a portion of its ancient character: the rest of the building is modernized into a spruce, comfortable residence, and is at this time occupied by a countryman of our own, General Hodgson.

The church at Creully is one of the most curious we have seen. The nave, side-aisles, and choir, are all purely Norman, except at the extremities. The piers are very massy; the arches wide and low; the capitals covered with rude, but most remarkable sculpture, which is varied on every pillar. Round the arches of the nave runs a band of the chevron ornament; and over them is a row of lancet windows, devoid of ornament, and sunk in a wall of extraordinary thickness. Externally, all is modernized.

The view of Caen, on entering from this direction, is still more advantageous than that on the approach from Lisieux. Time would not allow of our making any stop at the town on our return: we therefore proceeded immediately to Falaise, passing again through an open and monotonous country, which, thoughtfully cultivated, has a most dreary aspect from the scantiness of its population. We saw, indeed, as we went along, distant villages, thinly scattered, in the landscape, but no other traces of habitations; and we proceeded upwards of five leagues on our way, before we arrived at a single house by the road-side.

Falaise appeared but the more beautiful, from the impression which the desolate scenery of the previous country had left upon our minds. The contrast was almost equally pleasing and equally striking, as when, in travelling through Derbyshire, after having passed a tract of dreary moors, that seems to lengthen as you go, you suddenly descend into the lovely vallies of Matlock or of Dovedale. Not that the vale of Falaise may compete with those of Derbyshire, for picturesque beauty or bold romantic character; but it has features exclusively its own; and its deficiency in natural advantages is in some measure compensated, by the accessories bestowed by art. The valley is fertile and well wooded: the town itself, embosomed within rows of lofty elms, stretches along the top of a steep rocky ridge, which rises abrupt from the vale below, presenting an extensive line of buildings, mixed with trees, flanked towards the east by the venerable remains of the castle of the Norman Dukes, and at the opposite extremity, by the church of the suburb of Guibray, planted upon an eminence. Near the centre stands the principal church of Falaise, that of St. Gervais; and in front of the whole extends the long line of the town walls, varied with towers, and approached by a mound across the valley, which, as at Edinburgh, holds the place of a bridge.

The name *Falaise,* denotes the position of the town: it is said to be a word of Celtic origin; but I should rather suppose it to be derived from the Saxon, and to be a modification of the German word, *fels,* a rock, in which conjecture I find I am borne out by Adelung: *falesia,* in modern Latinity, and *falaise,* in French, signify a rocky shore. Hence, Brito, at the commencement of his relation of the siege by Philip Augustus, says,

"Vicus erat scabrâ circumdatus undique rupe,
Ipsius asperitate loci Falæsa vocatus,
Normannæ in medio regionis, cujus in altâ
Turres rupe sedent et mœnia; sic ut ad illam
Jactus nemo putet aliquos contingere posse."—

The dungeon of Falaise, one of the proudest relics of Norman antiquity, is situated on a very bold and lofty rock, broken into fantastic and singular masses, and covered with luxuriant vegetation. The keep which towers above it is of excellent masonry: the stones are accurately squared, and put together with great neatness, and the joints are small; and the arches are turned clearly and distinctly, with the key-stone or wedge accurately placed in all of them. Some parts of the wall, towards the interior ballium, are not built of squared free-stone; but of the dark stone of the country, disposed in a zigzag, or as it is more commonly called, in a herring-bone direction, with a great deal of mortar in the interstices: the buttresses, or rather piers, are of small projection, but great width. The upper story, destroyed about forty years since, was of a different style of architecture. According to an old print, it terminated with a large battlement, and bartizan towers at the angles. This dungeon was formerly divided into several apartments; in one of the lower of which was found, about half a century ago, a very ancient tomb, of good workmanship, ornamented with a sphynx at each end, but bearing no inscription whatever. Common report ascribed the coffin to Talbot, who was for many years governor of the castle; and at length an individual engraved upon it an epitaph to his honor; but the fraud was discovered, and the sarcophagus put aside, as of no account. The second, or principal, story of the keep, now forms a single square room, about fifty feet wide, lighted by circular-headed windows, each divided into two by a short and massy central pillar, whose capital is altogether Norman. On one of the capitals is sculptured a child leading a lamb, a representation, as it is foolishly said, of the Conqueror, whom tradition alleges to have been born in the apartment to which this window belonged: another pillar has an elegant capital, composed of interlaced bands.

Connected with the dungeon by a stone staircase is a small apartment, very much dilapidated, but still retaining a portion of its original facing of Caen stone. It was from the window of this apartment, as the story commonly goes, that Duke Robert first saw the beautiful Arlette, drawing water from the streamlet below, and was enamoured of her charms, and took her to his bed.—According to another version of the tale, the earliest interview between the prince and his fair mistress, took place as Robert was returning from the chace, with his mind full of anger against the inhabitants of Falaise, for having presumed to kill the deer which he had commanded should be preserved for his royal pastime. In this offence the curriers of the town had borne the principal share, and they were therefore principally marked out for punishment. But, fortunately for them, Arlette, the daughter of one Verpray, the most culpable of the number, met the offended Duke while riding through the street, and with her beauty so fascinated him, that she not only obtained the pardon of her father and his associates, but became his mistress, and continued so as long as he lived. From her, if we may give credence to the old chroniclers, is derived our English word, harlot. The fruit of their union was William the Conqueror, whose illegitimate birth, and the low extraction of his mother, served on more than one occasion as a pretext for conspiracies against his throne, and were frequently the subject of personal mortification to himself.—The walls in this part of the castle are from eight to nine feet thick. A portion of them has been hollowed out, so as to form a couple of small rooms. The old door-way of the keep is at the angle; the returns are reeded, ending in a square impost; the arch above is destroyed.

Talbot's tower, thus called for having been built by that general, in 1430 and the two subsequent years, is connected with the keep by means, of a long passage with lancet windows, that widen greatly inwards. It is more than one hundred feet high, and is a beautiful piece of masonry, as perfect, apparently, as on the day when it was erected, and as firm as the rock on which it stands. This tower is ascended by a staircase concealed within the substance of the walls, whose thickness is full fifteen feet towards the base, and does not decrease more than three feet near the summit. Another aperture in them serves for a well, which thus communicates with every apartment in the tower. Most of the arches in this tower have circular heads: the windows are square.—The walls and towers which encircle the keep are of much later date; the principal gate-way is pointed. Immediately on entering, is seen the very ancient chapel, dedicated to St. Priscus or, as he is called in French, St. Prix. The east end with three circular-headed windows retains its original lines: the masonry is firm and good. Fantastic corbels surround the summit of the lateral walls. Within, a semi-circular arch resting upon short pillars with sculptured capitals, divides the choir from the nave. In other respects the building has been much altered.—Henry Vth repaired it in 1418, and it has been since dilapidated and restored.—A pile of buildings beyond, wholly modern in the exterior, is now inhabited as a seminary or college. There are some circular arches within, which shew that these buildings belonged to the original structure.

Altogether the castle is a noble ruin. Though the keep is destitute of the enrichments of Norwich or Castle Rising, it possesses an impressive character of strength, which is much increased by the extraordinary freshness of the masonry. The fosses of the castle; are planted with lofty trees, which shade and intermingle with the towers and ramparts, and on every side they groupe themselves with picturesque beauty. It is said that the municipality intend to restore Talbot's tower and the keep, by replacing the demolished battlements; but I should hope that no other repairs may take place, except such as may be necessary for the preservation of the edifice; and I do not think it needs any, except the insertion of clamps in the central columns of two of the windows which are much shattered[94].

From the summit we enjoyed a delightful prospect: at our feet lay the town of Falaise, so full of trees, that it seemed almost to deserve the character, given by old Fuller to Norwich, of *rus in urbe*: the distant country presented an undulating outline, agreeably diversified with woods and corn-fields, and spotted with gentlemen's seats; while within a very short distance to the west, rose another ridgy mass of bare brown rock, known by the name of Mont Mirat, and still retaining a portion of the intrenchments, raised by our countrymen when they besieged Falaise, in 1417.—By this eminence the castle is completely commanded, and it is not easy to understand how the fortress could be a tenable position; as the garrison who manned the battlements of the dungeon and Talbot's tower, must have been exposed to the missiles discharged from the catapults and balistas planted on Mont Mirat.

The history of the castle is inseparably connected with that of the town: its origin may safely be referred to remote antiquity, the time, most probably, of the earliest Norman Dukes. If, however, we could agree with the fanciful author just quoted, it would claim a much earlier date. The very fact of its having a dungeon-tower, he maintains to be a proof of its having been erected by Julius Cæsar inasmuch as the word, dungeon, or, as it is written in French, donjon, is nothing but a corruption of Domus Julii! More than once in the course of this correspondence, I have called your attention to the fancies, or, to speak in plain terms, the absurdities, of theoretical antiquaries. The worthy priest, to whom we are indebted for the Recherches Historiques sur Falaise, "out-herods Herod." Writers of this description are curious and amusing, let their theories but rest upon the basis of fair probability. Even when we reject their reasonings, we are pleased with their ingenuity; and they serve, to borrow an expression from Horace, "the purpose of a whetstone." But M. Langevin has nothing farther to offer, than gratuitous assertion or vague conjecture; and yet, upon the faith of these, he insists upon our believing, that the foundation of Falaise took place very shortly after the deluge; that its name is derived from Felé, the cat of Diana, or from the less pure source of Phaloi-Isis; that the present site of the castle was that of a temple, dedicated to Belenus and Abraxas; and that every stone of remarkable form in the neighborhood, was either so shapened by the Druids, (notwithstanding it is the character of rocks, like those at Falaise, to assume fantastic figures,) or was at least appropriated by the Celtic priesthood to typify the sun, or moon, or stars.

Various tombs, stone-hatchets, &c., have been dug up at Tassilly, a village within six miles of Falaise, and fragments of mosaic pavements have been discovered in the immediate vicinity of the castle[95]; but history and tradition are alike silent as to the origin of these remains.—The first historical mention of Falaise is in the year 1027; during the reign of the fifth Norman Duke, Richard IIIrd, at which period this town was one of the strong holds of the duchy, and afforded shelter to Robert, the father of the Conqueror, when he rebelled against his elder brother. Falaise on that occasion sustained the first of the nine sieges, by which it has procured celebrity in history.—Fourteen years only elapsed before it was exposed to a second, through the perfidy of Toustain de Goz, Count of Hiesmes, who had been intrusted with the charge of the castle, and who, upon finding that his own district was ravaged by the forces of the King of France, voluntarily offered to surrender to that monarch the fortress under his command, on condition that his territory, the Hiesmois, should be spared. But Duke William succeeded in retaking the place of his birth before the traitor had an opportunity of introducing the troops of his new ally.—In the years 1106 and 1139, Falaise opposed a successful resistance to the armies of Henry Ist, and of Geoffrey Plantagenet. Upon the first of these occasions, the Count of Maine, the general of the English forces, retired with shame from before the walls; and Henry was foiled in all his attempts to gain possession of the castle, till the battle of Tinchbray had invested him with the ducal mantle, and had induced Robert himself to deliver up the fortress in person to his more fortunate brother. On the second occasion, Robert Marmion, lord of the neighboring barony of Marmion le Fontenay, a name equally illustrious in Norman and in English story, held Falaise for Eustace of Boulogne, son to Stephen, and twice repelled the attacks of the husband of the Empress Maud.—The fourth siege was conducted with different success, by Philip Augustus: for seven days the citizens quietly witnessed the preparations of the French monarch; and then, either alarmed by the impending conflict, or disgusted by the conduct of their own sovereign, who had utterly deserted them, they opened their gates to the enemy.—In 1417 the case was far otherwise, though the result was the same. Henry Vth attacked Falaise upon the fourth of November, and continued to cannonade it till the middle of the following February; and, even then, the surrender was attributed principally to famine. Great injuries were sustained by the town in the course of this long siege; but, to the credit of our countrymen, the efforts made towards the reparation of them were at least proportionate. The fortifications were carefully restored; the chapel was rebuilt and endowed afresh; Talbot's tower was added to the keep; and a suite of apartments, also named after that great captain, was erected in the castle.—The resistance made by the English garrison of Falaise in 1450, at the time when we were finally expelled from the duchy, was far from equal to that which the French, had previously shewn. Vigour was indeed displayed in repeated sallies, but six days sufficed to put the French general in possession of the place. Disheartened troops, cooped up in a fortress without hope of succour, offer but faint opposition; and Falaise was then the last place which held out in Normandy, excepting, only Domfront and Cherbourg, both which were taken almost immediately afterwards.—Falaise, from this time forwards, suffered no more from foreign enemies: the future miseries of the town were inflicted by the hands of its own countrymen. In common with many other places in France, it was doomed to learn from hard experience, that "alta sedent civilis vulnera dextræ."—Instigated by the Count de Brissac, governor of the town, and one of the most able generals of the league, the inhabitants were immoveable in their determination to resist the introduction of tenets which they regarded as a fatal variance from the Catholic faith. The troops of Henry IIIrd, in alliance with those of his more illustrious successor, were vainly brought against Falaise in 1589, by the Duc de Montpensier; a party of enthusiastic peasants, called Gautiers, from the name of a neighboring village, where their association originated, harassed the assailants unremittingly, and rendered such effectual assistance to the garrison, that the siege was obliged to be raised.—But it was only raised to be renewed at the conclusion of the same year, by Henry of Bourbon, in person, whom the tragical end of his late ally had placed upon the throne of France. Brissac had now a different enemy to deal with: he answered the king's summons to surrender, by pleading his oath taken upon the holy sacrament to the contrary; and he added that, if it should ultimately prove necessary for him to enter into any negotiation, he would at least delay it for six months to come. "Then, by heavens!" replied Henry, "I will change his months into days, and grant him absolution;" and; so saying, he commenced a furious cannonade, which soon caused a breach, and, in seven days, he carried the town by assault. Brissac, who, on the capture of the fortress, had retired into the keep, found himself shortly afterwards obliged to capitulate; and I am sorry to add, that the terms which he proposed and obtained, were not of a nature to be honorable to his character. The security of his own life and of that of seven of his party, was the principal stipulation in the articles. The rest of the garrison were abandoned to the mercy of the conqueror, who contented himself with hanging seven of them in memorial of the seven days of the siege; but, if we may believe the French historians, always zealous for the honor of their monarchs, and especially of this monarch, Henry selected the sufferers from among those, who, for their crimes, had, subjected themselves to the pain of death.

From these various attacks, but principally from those of 1417 and 1589, the fortifications of Falaise have suffered materially; and since the last no care has been taken to repair them. The injuries sustained at that period, and the more fatal, though less obvious ones, wrought by the silent operation of two centuries of neglect, have brought the walls and towers to their present state of dilapidation.

The people of Falaise are commonly supposed to be Normans *κατ εξοχην* [English. Not in Original: pre-eminently, especially, above all]; and when a Norman is introduced upon the French stage, he calls himself a Falesian, just as any Irishman, in an English farce, is presumed to come from Tipperary. The town in the French royal calendar is stated to contain about fourteen thousand inhabitants; but we are assured that the real number does not exceed nine thousand. Its staple trade is the manufacture of stockings, coarse caps, and lace. The streets are wide; and the public fountains, which are continually playing, impart a freshness, which, at the present burning season, is particularly agreeable.—The town now retains only four churches, two within its precincts, and two in the suburbs. The revolution has deprived it of eight others. Of those which are now standing, the most ancient is that situated near the castle, and dedicated to the Holy Trinity. Langevin assures us that it was built upon the ruins of the temple of Felé, Isis, Belenus, and the heavenly host of constellations, and that in the fifth century it changed its heathen for its Christian patrons. The oldest part (a very small one it is) of the present structure, appertains to a building which was consecrated in 1126, by the Archbishop of Rouen, in the presence of Henry Ist, but which was almost entirely destroyed by the cannonade in the fifteenth century. An inscription in gothic letters, near the entrance, relates, that after this desolation, a beginning was made towards the re-building of the church, "in 1438, a year of war, and death, and plague, and famine;" but it is certain that not much of the part now standing can be referred even to that period. The choir was not completed till the middle of the sixteenth century, nor the Lady-Chapel till the beginning of the following one. Architecturally considered, therefore, the church is a medley of various styles and ages.

The larger church, that of St. Gervais and St. Protais, is said to have been originally the ducal chapel, and to stand in the immediate vicinity of the site of the Conqueror's palace, now utterly destroyed. According to an ancient manuscript, this church was consecrated at the same time as that of the Trinity. The intersecting circular-headed arches of its tower are curious. The Norman corbel-table and clerestory windows still remain; and the exterior of the whole edifice promises a gratification to a lover of architectural antiquity, which the inside is little calculated to realize.—An invading army ruined the church of the Trinity; civil discord did the same for that of St. Gervais. The Huguenots, not content with plundering the treasure, actually set fire to the building, and well nigh consumed it: hence, the choir is the work of the year 1580, and the southern wall of the nave is a more recent construction.

We see Falaise to a great advantage: every inn is crowded; every shop is decked out; and the streets are full of life and activity; all in preparation for the fair, which commences in three days, on the fifteenth of this month, the anniversary of the Assumption of the Holy Virgin. This fair, which is considered second to no other in France, excepting that of Beaucaire, is held in the suburbs of Guibray, and takes its name from the place where it is held. For the institution, Falaise is indebted to William the Conqueror; and from it the place derives the greatest share of its prosperity and importance. During the fourteen days that the fair continues, the town is filled with the neighboring gentry, as well as with merchants and tradesmen of every description, not only from the cities of Normandy, but from Paris and the distant provinces, and even from foreign countries. The revolution itself respected the immunities granted to the fair of Guibray, without, at the same time, having the slightest regard, either to its royal founder, or its religious origin.—An image of the Virgin, discovered under-ground by the scratching and bleating of a lamb, first gave the stamp of sanctity to Guibray. Miraculous means had been employed for the discovery of this statue; miraculous powers were sure to be seated in the image. Pilgrims crowded from all places to witness and to adore; and hawkers, and pedlars, and, as I have seen inscribed upon a hand-bill at Paris, "the makers of he-saints and of she-saints," found Guibray a place of lucrative resort. Their numbers annually increased, and thus the fair originated.—We are compelled to hasten, or we would have stopped to have witnessed the ceremonies, and joined the festivities on the occasion. Already more than one field is covered with temporary buildings, each distinguished by a flag, bearing the name and trade of the occupant; already, too, the mountebanks and showmen have taken their stand for the amusement of the company, and the relaxation of the traders; and, what is a necessary consequence of such assemblages, you cannot stir without being pestered with crowds of boys, proffering their services to transport your wares.

The church of Guibray, like the others of Falaise, offers specimens of Norman architecture, strangely altered and half concealed by modern innovations. In the first syllable of the name of the place, you will observe the French word for misletoe, and may thence infer, and probably not without reason, the antiquity of the station; the latter syllable, albeit in England sheep are not wont to *bray*, is supposed by the pious to have reference to the bleating of the lamb, which led to the discovery of the miraculous image.—Etymology is a wide district in a pleasant country, strangely intersected by many and deceitful paths. He that ventures upon the exploring of it, requires the utmost caution, and the constant control of sober reason: woe will be sure to betide the unfortunate wight, who, in such a situation, gives the reins to fancy, and suffers imagination to usurp the place of judgment, without reflecting, as has been observed by the poet on a somewhat similar occasion, that

> "Tis more to curb than urge the generous steed,
> Restrain his fury, than provoke his speed."

Footnotes:

[94] The outline of the castle is egg-shaped; and the following are its dimensions, in French measure, according to M. Langevin.—Length, 720 feet; mean width, 420; quantity of ground contained within the walls, two acres and a perch.

[95] Recherches Historiques sur Falaise, p. XIX. and XXIX.

LETTER XXX
ROCK AND CHAPEL OF ST. ADRIEN—PONT-DE-L'ARCHE—PRIORY OF THE TWO LOVERS—ABBEY OF BONPORT—LOUVIERS—GAILLON—VERNON

(*Mantes, August*, 1818)

The last letter which I wrote to you, was dated from Falaise. Look in the map and you will see that you now receive one from a point completely opposite. In four days we have passed from one of the most western towns of the province, to a place situated beyond its eastern frontier; and in four more, we may almost hope to be with you again. In this hasty journey we travelled through a district which has not yet become the subject of description to you; and though we travelled with less comfort of mind, than in the early part of our tour, I am yet enabled to send you a few details respecting it.

From Falaise we went in a direct line to Croissanville: the road, which we intended to take by St. Pierre sur Dive to Lisieux, was utterly impracticable for carriages. From Croissanville to Rouen we almost retraced our former steps: we did not indeed again make a *détour* by Bernay; but the straight road from Lisieux to Brionne is altogether without interest.

There are two ways from Rouen to Paris: the upper, through Ecouis, Magny, and Pontoise; the lower, by the banks of the Seine. Having travelled by both of them before, we could appreciate their respective advantages; and we knew that the only recommendation of the former was, that it saved some few miles in distance; while the latter is one of the most beautiful rides in France, and the towns, through which it passes, are far from being among the least interesting in Normandy. In such an alternative, there was no difficulty in fixing our choice, and we proceeded straight for Pont-de-l 'Arche. The chalk cliffs, which bounded the road on our left, for some distance from Rouen, break near the small village of Port St. Ouen, into wild forms, and in one spot project boldly, assuming the shape of distinct towers. These projections are known by the name of the rock of St. Adrien; thus called from the patron saint of a romantic chapel, a place of great sanctity, and of frequent resort with pilgrims, situated nearly mid-way up the cliff.—The chapel is indeed little more than an excavation, and is altogether so rude, that its workmanship affords no clue to discover the date of the building. Its south side and roof are merely formed of the bare rock. To the north it is screened by an erection, which, were it not for the windows and short square steeple, might easily be mistaken for a pent-house. The western end appears to display some traces of Norman architecture. The hill, which leads to this chapel, commands a view of Rouen, the most picturesque, I think, of all that we have seen of this city, so picturesque from various points. You can scarcely conceive the eagerness with which we endeavored to catch the last glimpse, as the prospect gradually vanished from our sight, or the pleasure with which we still dwell, and shall long continue so to do, upon the recollection. All round the chapel, the bare chalk is at this time tinged with a beautiful glow, from the blue flowers of the Viola Rothomagensis: the Isatis Tinctoria, the true Woad, is also common on the steep sides of the cliff. This plant, which is here indigenous, became, during the reign of Napoléon, an object of attention with the government, as a succedaneum for indigo, at the same time that beet-root was destined to supply the continent with sugar, and salsafy, or parched wheat, to hold the place of coffee. The restoration of peace has caused the Isatis to be again neglected; but the Reseda luteola, or, Dyer 's woad, is much cultivated in the neighborhood, as is the Teasel for the use of the cloth manufactory.

Pont-de-l 'Arche, though now a small mean town, may boast of high antiquity, if it be rightly believed to be the ancient Pistae, the seat of the palace erected by Charles the Bald, in which that sovereign convened councils in the years 861 and 869, and held assemblies of his nobles in 862 and 864; and from which, his edicts promulgated in those years, are dated. The same monarch also built here a magnificent bridge, defended at one extremity by a citadel upon a small island.—From this there seems every reason to believe that the town has derived its name; for, in a diploma issued by our Henry IInd, he calls the place Pontem Arcis; and its present appellation is nothing but its Latin name translated into French. The fortress at the head of the bridge was demolished about thirty years ago, at the time when Millin published his[96] account of the town. The plate attached to that account, represents one of the towers as still standing.—Though deprived of its citadel, Pont-de-l'Arche retains to the present day its walls, flanked by circular towers; and its bridge, which is the lowest stone bridge down the Seine, is a noble one of twenty-two arches, through which the river at a considerable depth below, rolls with extraordinary rapidity. In the length of this bridge are some mills, which are turned by the stream; and the current is moderated under one of the arches, by a lock placed on the down-stream side, into which barges pass, and so proceed with security; The bridge, with its mills, forms a very picturesque object.

At a short distance from the bridge, to the left, looking towards Paris, is the Colline des deux amans, formerly surmounted by the priory of the same name. Of the history of the monastery nothing is known with certainty, nor is even the date of its foundation ascertained, though it is stated by Millin to be one of the most ancient in Normandy[97]. But the traditionary tale connected with this convent, forms the subject of one of the lays of Mary of France; and it has been elegantly translated by the late Mr. Ellis, in the introduction to his History of our Ancient Metrical Romances;—Du Plessis[98] is, however, of opinion, that the name of the priory is nothing more than a corruption from the words, deux monts, in allusion to the twin hills, on one of which it stands; or, if lovers must have any thing to do with the appellation, he piously suggests that divine love may have been intended, and that the parties were no other than our Savior and the Virgin, whose images were placed over the door of the conventual church.

On the opposite side of the bridge of Pont-de-l'Arche, stand the remains of a far richer abbey, that of Bonport, of the Cistertian order, founded by Richard Coeur-de-Lion, in 1190, as an ex voto. The monarch, then just in possession of his crown, was indulging with his courtiers in the pleasures of the chace, and, carried away by the natural impetuosity of his temper, had plunged in pursuit of the deer into the Seine, whose rapid current brought his life into imminent danger; and he accordingly vowed, if he escaped with safety, to erect a monastery upon the spot where he should reach the shore. Hence, according to Le Brasseur[99], the foundation, and hence the name. I ought, however, to add, that no record of the kind is preserved in the Neustrta Pia, nor even by Millin, who has described and figured such of the monastic buildings and monuments as had been spared at the early part of the revolution[100]. Another view of the ruins has since been published by Langlois, in the first number of a work which was intended to have comprised a long series of Norman antiquities, but was discontinued for want of encouragement. The author, whose portrait I have sent you in the course of this correspondence, is himself a native of Pont-de-l'Arche, and has subjoined to his fas-ciculus a couple of plates, illustrative of the costume and customs of the neighborhood.—In one of these plates, an itinerant male fortune-teller is satisfying a young peasant as to the probability of her speedy marriage, by means of a pack of cards, from which he has turned up the king and queen and ace of hearts. In the other, a cunning woman is solving a question by a book and key. The poor girl's sweetheart is an absent soldier, and fears and doubts are naturally entertained for his safety. To unlock the mysteries of fate, the key is attached to the mass-book, and suspended from the tip of the finger of the sybil, who reads the first chapter of the gospel of St. John; and the invocation is answered by the key turning of its own accord, when she arrives at the verse beginning, "and the word was made flesh[101]."—A fine rose-window in the church of the abbey of Bonport, and two specimens of painted glass from its windows, the one representing angels holding musical instruments, supposed to be of the thirteenth century, the other containing a set of male and female heads of extraordinarily rich color, probably executed about a century later, are given by Willemin in his very beautiful Monumens Français inédits. In the same work, you will likewise find two still more interesting painted windows from Pont-de-l'Arche; some boatmen and their wives in the Norman costume of the end of the sixteenth century, and a citizen of the town with his lady, praying before a fald-stool, bearing the date, 1621.

The church of Pont-de-l'Arche, though greatly dilapidated, is a building worth notice, in a fine style of the decorated gothic. The nave is very lofty; the high altar richly carved and gilt; the oak pulpit embossed with saints; and the font covered with curious, though not ancient, sculpture. Rich tracery abounds in the windows, which are also filled with painted glass, some of it of very good quality. Scripture history and personages occupy, as usual, the principal part; but in one of the windows we noticed a representation of the Seine full of islands, and the town of Pont-de-l'Arche, with a number of persons quitting it with their horses, baggage, &c. in apparent confusion. So shattered, however, is the window, that the story is no longer intelligible in its details; and fragments, quite illegible, are all that remain of the inscriptions formerly beneath it. It is probable, that the intention of the artist was to give a picture of the miseries experienced by the inhabitants at the burning of the town by our troops under Edward IIIrd.—On the south side of the church the buttresses are enriched with canopies and other sculpture; and there was originally a highly-wrought balustrade, ornamented with figures of children, a part of which remains.—Pont-de-l'Arche claims the merit of having been the first town in France, which acknowledged Henry IVth as its lawful sovereign, after the assassination of his predecessor, in 1589.

On leaving this place, we passed through the forest of the same name, an extensive tract covered with young trees, principally beech, oak, and birch. The soil, a mixture of chalk and gravel, is poor, and offers but little encouragement to the labors of the plough. All around us, the distant prospect was pleasantly varied with gentle hills, upon one of which, nearly in front, we soon saw Louviers, a busy manufacturing town, of about seven thousand inhabitants, who are chiefly employed in making the fine cloth of the district, which is considered superior in quality to any other in France. Spanish wool is almost exclusively used for the purpose.

Throughout the vicinity of Louviers, are the most undoubted symptoms of commercial prosperity; new houses every where erecting, and old ones undergoing improvement. But the streets of the town itself are, as usual, dirty and narrow, and the people of the lower orders more than commonly ragged and beggarly. It was impossible to mistake the nature of their occupations; so many of them had their faces and hands, and every part of their limbs and bodies that was visible, died of a bright blue.—The church at Louviers is very much injured, but very handsome; and though reduced to a nave with its four aisles it is still a spacious edifice. The south porch, which projects boldly in the form of a galilee, is scarcely to be excelled as a specimen of pointed architecture at its highest pitch of luxuriant beauty. Yet, even in this, the saints have been torn from their pedestals by the wanton violence of the Calvinists or democrats. The central tower is square and short: it is, however, handsome. Two windows, very similar to those of the tower of St. Romain, in Rouen cathedral, light it on either side; and saints, placed under canopies, ornament the angles behind the buttresses.—The great western door is closed, and the front defaced: the eastern end, likewise, is altogether modern.—Within, the same kind of architecture prevails as in the exterior, but the whole is so concealed, and degraded by ornaments in the worst of taste, and by painted saints in the most tawdry dresses, that the effect is disgusting. I never saw so great an array of wretched representations of the heavenly host: the stone images collected round the holy sepulchre, are even worse than those at Dieppe. Near the chapel of the sepulchre, however, are four bas-reliefs, attached to the wall, exhibiting different events in our Savior's life of good execution, and not in had taste: an open gallery of fillagree stone-work, under the central tower on the south side, is an object really deserving of admiration.

M. Langlois has engraved the gable end of an old house at Louviers, said to have belonged to the Knights Templars. We found it used as an engine-maker's shop; and neither within nor without, could we discover any thing to justify his opinion, that it is a building of the twelfth or thirteenth century. On the contrary, the windows, which are double, under a flatly-pointed arch, and are all of them trefoil-headed, would rather cause it to be considered as erected two centuries later.

The town of Louviers, though never fortified, is noticed on several occasions in history. It was the seat of the conferences between Richard Coeur-de-Lion and Philip Augustus, which ended in the treaty of 1195, defining new limits to Normandy.—It was, as I have already mentioned, one of the items of the compensation made by the same Duke to the Archbishop of Rouen, for the injury done to the church, by the erection of Château Gaillard.—During the wars of Edward IIIrd, "Louviers," to use the language of old Froissart, "after the battle of Caen, was soon entered by the Englishmen, as it was not closed; and they over-ran, and spoiled, and robbed it without mercy, and won great riches; for it was the chief place in all Normandy for drapery, and was full of merchandize."—And, in the subsequent warfare of the fifteenth century, this town, like the others in the duchy, was taken by our countrymen, under Henry Vth, and lost by them under his successor.—Hither the Norman parliament retired when the Huguenots were in possession of Rouen; and here they remained till the recapture of the capital.—It was probably owing in a great measure to this circumstance, that Louviers was induced to distinguish itself by a devoted attachment to the party of the league, for which it suffered severely in 1591, when it was captured and pillaged by the royalists shortly after their victory at Ivry. The town was then taken through the treachery of a priest of the name of Jean de la Tour, who received, as a recompence, a stall in the cathedral at Evreux, but was so much an object of abhorrence with his brethren, that he scarcely ever ventured to appear in his place. During the holy week, however, he attended; and it once happened, that while he was so officiating, all the canons contrived to leave the church towards the close of the psalm, which immediately precedes the Benedictus at Laudes, so that the anthem, Traditor autem, which is sung with that hymn, necessarily fell to the part of de la Tour, who found himself compelled to chaunt it, to his own extreme confusion, and the infinite amusement of the congregation. Irritated and mortified, the poor priest preferred his complaints to the king; but it was one thing to love the treason, and another to love the traitor; and his appeal obtained no redress.

From Louviers our next stage was Gaillon, on our road to which we passed some vineyards, the most northern, I believe, in Normandy. The vines cultivated in them are all of the small black cluster grape; and the wine they produce, I am told, is of very inferior quality,—No place can appear at present more poverty-stricken than Gaillon; but the case was far otherwise before the glories of royal and ecclesiastical France were shorn by the revolution. Ducarel, who visited this town about the year 1760, dwells with great pleasure upon the magnificence of its palace and its Carthusian convent and church. Of the palace the remains are still considerable; and, after having been suffered to lie in a state of ruin and neglect from an early period in the revolution, they are now fitting up as a prison. The long inscription formerly over the gate might with great propriety be replaced by the hacknied phrase, "Sic transit gloria mundi;" for the vicissitudes of the fortune of noble buildings are strikingly illustrated by the changes experienced by this sumptuous edifice, long proverbial throughput France for its splendor.

Philip Augustus conferred the lordship of Gaillon upon one of his captains of the name of Cadoc, as a reward for his activity in the conquest of Normandy. Louis IXth afterwards, early in the thirteenth century, ceded the town in perpetuity to the Archbishop of Rouen. St. Louis here received by way of exchange the Château of Pinterville, which he bestowed upon William d'Aubergenville, whose uncle, the Bishop of Evreux, had, while chancellor of France, done much service to him and to Queen Blanche, his mother. From that time to the revolution the archbishops had their country seat at Gaillon, and enjoyed the sole right of trying civil and criminal causes within the town and its liberties. Their palace, which was destroyed during the wars of Henry Vth, in 1423, was rebuilt about a century afterwards by the munificence of the first cardinal Georges d'Amboise, one of whose successors in the prelacy, Colbert, expended, as it is said, more than one hundred thousand livres towards the embellishment of it.—Another archbishop, the Cardinal of Bourbon, founded the neighboring monastery, in the year 1571. The conventual church was destroyed by fire, through the carelessness of some plumbers, shortly after Ducarel visited it; and with it perished the celebrated monument of one of the counts of Bourbon Soissons, said to have been a master-piece of sculpture.

The limits assigned to Normandy by the treaty of Louviers, made Gaillon a frontier town of the duchy; and here therefore I should take my leave of you, but that, in the prouder days of its history, Vernon was likewise swayed by the ducal sceptre. Vernon also seems peculiarly connected with England, from the noble family of the same name still flourishing, agreeably to their well-known punning motto, on your side of the water. This motto is in the highest degree inapplicable to the present state of the town, whose old and ruinous appearance looks as if it had known neither improvement nor repair for centuries. Better things might have been expected from the situation of Vernon, on the banks of the Seine, in a singularly beautiful valley, and from its climate, which is reported to be so extraordinarily healthy, that instances of individuals attaining in it the age of one hundred are not unfrequent.

The royal palace, formerly here, is now wholly swept away; and of the ancient fortifications there remains little more than a tower, remarkable for the height and thickness of its walls, a part of the castle, which, in the reign of Henry IInd, was held by the service of sixteen knights for its defence[102].—Prior to the revolution, Vernon contained five religious houses, three of them founded by St. Louis, who is said to have regarded this town with peculiar favor, and probably on that account assigned it as a jointure to his queen, an honor which it has received upon more than one other occasion.

The present parish church of Vernon was collegiate. It was founded about the year 1052, by William of Vernon, and was endowed by him, at the time of its dedication, with the property called, La Couture du Pré de Giverny, and with a fourth part of the forest of Vernon, all which the dean and canons continued to enjoy till the revolution. This William appears to have been the first of the family who adopted the surname of Vernon. His son, Richard, by whom the foundation was formally confirmed, attended the Conqueror to England, and obtained there considerable grants. One of their descendants ceded the town in 1190 to the King of France, accepting in return other lands, according to a treaty still preserved in the royal library at Paris. The tombs of the founder, and of his namesake, Sir William de Vernon, constable of England, who died in 1467, and of many others of the family, among the rest the stately mausoleum of the Maréchal de Belle Isle, were destroyed during the reign of jacobinism and terror. The portraits, however, of the Marshal and of the Duc de Penthièvre, both of them very indifferent performances, were saved, and are now kept in the sacristy. The only monument left to the church is that of Marie Maignard, whose husband, Charles Maignard, was Lord of Bernières and president of the parliament of Normandy. She died in 1610. Her effigy in white marble, praying before a fald-stool, has also been spared.

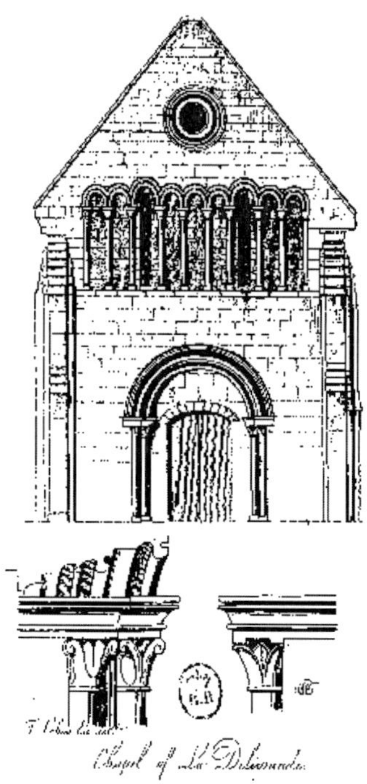

The church itself is a spacious building, consisting of a nave and two aisles, with chapels beyond, separated by lofty pointed arches, supported on clustered pillars, to each of which is still attached a tabernacle; but the statues have been destroyed. The choir is altogether in a different style of architecture: that portion of it which immediately surrounds the altar, is early Norman, and most probably belonged to the original structure. Its arches vary remarkably in width. The most narrow among them are more decidedly horseshoe-shaped, than any others which I recollect to have seen.—The west front, though much mutilated, is still handsome. It is flanked by two small, very short turrets, richly ornamented.—The square central tower, capped by a conical roof, does not even equal the height of the nave, which is greatly superior to that of the choir.—Upon an eminence in the immediate vicinity of Vernon, are the remains of a Roman encampment.

With Vernon we quitted ancient Normandy: our ride thence to Mantes has been delightful; and this town, for the excellence of its buildings, for neatness, and for a general air of comfort, far excels any other which we have seen in the north of France. The name of Mantes also recals the memory of the Duc de Sully, and recals that of the Conqueror, whose life fell a sacrifice to the barbarous outrage of which he was here guilty.—But, I now lay down my pen, and take my leave of Normandy, happy, if by my correspondence during this short tour, I have been able to impart to you a portion of the gratification which I have myself experienced, while tracing the ancient history, and surveying the monuments of that wonderful nation, who, issuing from the frozen regions of the north, here fixed the seat of their permanent government, became powerful rivals of the sovereigns of France, saw Sicily and the fairest portion of Italy subject to their sway, and, at the same time that they possessed themselves of our own island, by right of conquest, imported amongst us their customs, their arts, and their institutions, and laid the basis of that happy constitution, under which, by the blessing of God, Britain is at this moment the pride and envy of the world!

Footnotes:

[96] Antiquités Nationales, IV. No. 48.

[97] Antiquités Nationales, II. No. 17.

[98] Histoire de la Haute Normandie, II. p. 332.

[99] Histoire d'Evreux, p. 161.

[100] Antiquités Nationales, IV. No. 40.

[101] This mode of divination by the Bible and key, is also to be found among the superstitions of our own country.—See Ellis' edition of Brand's Popular Antiquities, II. p. 641.

[102] Ducarel's Anglo-Norman Antiquities, p. 93.—Respecting Vernon, see also Millin, Antiquités Nationales, III. No. 26, in which four plates, and near fifty pages of letter-press, are devoted to this town.

APPENDIX I

The printing of this work was just concluded, when the author was favored with drawings, accompanied with short descriptions, of the chapel of our *Lady of the Délivrande,* near Caen, and of an ancient font at Magneville, near Valognes. For the former he is indebted to Mr. Cohen, to whom he has so often in the course of the work, had occasion to express his obligations; for the latter, to M. de Gerville, an able antiquary at Valognes. Both these subjects are of such a nature, that he is peculiarly happy to be able to add them to his imperfect account of the Antiquities of Normandy: the whole duchy does not contain a religious building more celebrated for its sanctity than the chapel; and while ancient fonts of any description are rare in the province, he doubts if another is to be found like that of Magneville, ornamented with sculpture and an inscription.

Some historians suppose, that the country situated between Caen and the sea, formed at least, a part of the Saxon shore of Neustria. Amongst the other ancient buildings which are found in this district, the chapel of Notre Dame de la Délivrande, to which the Normans have resorted in pilgrimage during the last eight hundred years, is, perhaps, the most remarkable.

When the philosophers of the revolution envied the religious enjoyments of the common man, all pilgrimages were forbidden, and the road leading to our Lady's Chapel, and which, indeed, is the only high road in this part of the country, became almost impassable. Under the Emperor it was thoroughly repaired, and, as they say, by his especial order; and since the accession of the present French king, the fathers of the mission, who lose no favorable opportunity of fostering the spirit of devotion, have erected roods and tabernacles, at due distances, all along the way side.

After leaving Caen, the traveller will not fail to linger on the little hill which he ascends just after passing by the first crucifix. Hence he enjoys a lovely prospect, such as delighted the old masters. In the foreground is the lofty cross, standing on a quadrangular pyramid of steps. The broken hollow path bending upwards round the base, is always occupied by a grotesque group of cripples and beldames, in rags and tatters, laughing and whining and praying. The horizon is bounded by long lines of grey and purple hills, nearer are fields and pastures, whilst the river glitters and winds amidst their vivid tints. Nearer still, the city of Caen extends itself from side to side, terminated at each extremity by the venerable abbeys of William and Matilda. There are no traces of work-shops and manufactories, or of their pollution; but the churches with their towers and spires rise above the houses in bold architectural masses, and the city assumes a character of quiet monastic opulence, comforting the eye and the mind.

About four miles farther on from Caen, we reached Cambre, one of the many seignories which belonged to the very noble family of Mathan. There was a Serlo de Mathan, who appears as a witness to one of the Conqueror's charters, and the family is now represented by the present Marquis, who has recovered his château, and a fragment of his domain. Cambre is also the residence of the Abbé de la Rue, by whom the Marquis was educated. When they both took refuge in England, the Abbé was the only protector of his pupil, who now returns the honorable obligation. It is well known that the Abbé has devoted his life to the investigation of the antiquities both of Normandy and of the Anglo-Normans. Possessing in a high degree the acute and critical spirit of research which distinguished the French archaiologists of the Benedictine school, we have only to regret, that the greater part of his works yet remain in manuscript. His History of Anglo-Norman Poetry, which is quite ready for the press, would be an invaluable accession to our literature; but books of this nature are so little suited to the taste of the French public, that, as yet, he has not ventured upon its publication. The collections of the Abbé, as may be anticipated, are of great value; they relate almost wholly to the history of the duchy. The château escaped spoliation. The portraits of the whole line of the Mathans, from the first founder of the race, in his hauberk, down to the last Marquis, in his frisure, are in good preservation; and they are ancient specimens of the sign-post painting usually found in old galleries. The Marquis has also a finely-illuminated missal, which belonged to a Dame de Mathan, in the fourteenth century, and which has been carefully handed down in the family, from generation to generation.

The church of Douvre, the next village, is rather a picturesque building. The upper story of the tower has two pointed windows of the earliest date. A pediment between them rests on the archivolt on either side. This is frequently seen in buildings in the circular style. The other stories of the tower, and the west front of the church are Norman; the east end is in ruins.

The British name of the village may afford ground for much ethnigraphical and etymological speculation.

Saint Exuperius is said to have founded the Chapel of La Délivrande, some time in the first century. The tradition adds, that the chapel was ruined by the Northmen,—and the statue of the Virgin, which now commands the veneration of the faithful, remained buried until the appointed time of resuscitation, in the reign of Henry Ist, when it was discovered, in conformity to established usage and precedent in most cases of miraculous images, by a lamb. Baldwin, Count of the Bessin and Baron of Douvre, was owner of the flock to which the lamb belonged. The Virgin would not remain in the parish church of Douvre, in which she was lodged by the Baron, but she returned every night to the spot where she was disinterred. Baldwin therefore understood that it was his duty to erect a chapel for her reception, and he accordingly built that which is now standing, and made a donation of the edifice to the Bishop of Bayeux, whose successor receives the mass-pennies and oblations at this very day. Some idea of the architecture of the building may be formed from the inclosed sketch of the western front. During the morning mass, the chapel was crowded with women, young and old, who were singing the litany of the Virgin in a low and plantive tone. A hymn of praise was also chaunted. It was composed by the learned Bishop Huet, and it is inscribed upon a black marble tablet, which was placed in the chapel by his direction. The country women of the Saxon shore possess a very peculiar physiognomy, denoting that the race is unmixed. The Norman-Saxon damsel is full and well made, her complexion is very fair, she has light hair, long eyelashes, and tranquil placid features; her countenance has an air of sullen pouting tenderness, such as we often find in the women represented in the sculptures and paintings of the middle ages. And all the girls are so much alike, that it might have been supposed that they all were sisters. As to our Lady, she is gaily attired in a Cashemire shawl, and completely covered with glaring amber necklaces and beads, and ribband knots, and artificial flowers. Many votive offerings are affixed round her shrine. The pilgrim is particularly desired to notice a pair of crutches, which testify the cure of their former owner, who lately hobbled to the Virgin from Falaise, as a helpless cripple, and who quitted her in perfect health. Of course the Virgin has operated all the usual standard miracles, including one which may be suspected to be rather a work of supererogation, that of restoring speech to a matron who had lost her tongue, which had been cut out by her jealous husband. Miracles of every kind are very frequently performed, yet, if the truth must be told, they are worked, as it were, by deputy, for the real original Virgin suffered so much during the revolution, that it has been thought advisable to keep her in the sacristy, and the statue now seen is a restoration of recent workmanship. In order to conciliate the sailors and fishermen of the coast, the Virgin has entered into partnership with St. Nicholas, whose image is impressed on the reverse of the medal representing her, and which is sold to the pilgrims.

The country about La Délivrande is flat, but industriously cultivated and thickly peopled. The villages are numerous and substantial. From a point at the extremity of the green lane which leads onward from La Délivrande, six or eight church spires may be counted, all within a league's distance. By the advice of the Abbé de la Rue, we proceeded to Bernieres, which is close to the sea. The mayor of the commune offered his services with great civility, and accompanied us to the church, which, as he told us, was built by Duke William. We easily gave credit to the mayor's assertion, as the interior of the nave is good Norman. The pillars which support the groining of the roof are square; this feature is rather singular. The tower and spire are copied from Saint Peter, at Caen. Those of Luc, Courseilles, Langrune, and the other neighboring villages, are upon the same model. Many instances of the same kind of affiliation occur at home, which shew how easily a fashion was set in ecclesiastical architecture.

APPENDIX II

The most remarkable among the ancient inscriptions found in that part of Normandy, which is now comprised in the Department of La Manche, are upon an ancient altar, at Ham, on a medallion attached to the outside of the church of Ste. Croix, at St. Lô, and upon the font at Magneville, near Valognes. The first of these has generally been referred to the seventh century; the second seems to be of the ninth; and the last may with safety be considered as of the latter part of the tenth, or beginning of the eleventh, at which period, the choir of the church of Magneville appears also to have been erected. Of the sculpture upon the font, as well as of the inscription, an accurate idea may be formed, from the annexed drawing: the most remarkable character of the inscription seems to be in its punctuation. The letters upon the altar, at Ham, touch one another, and there is no separation of any kind between the words: here, on the contrary, almost all the words are divided by three or four points placed in a perpendicular direction, except at the end of the phrases, where stops are wholly wanting. At Ham, also, the letters are cut into the stone, while at Magneville they are drawn with a brush, with a kind of black pigment.

G.

INDEX